Marius Andrei Olariu

Fabrication of Flexible Screen-Printed Dielectrophoretic Devices

Marius Andrei Olariu

Fabrication of Flexible Screen-Printed Dielectrophoretic Devices

LAP LAMBERT Academic Publishing

Publisher:
LAP LAMBERT Academic Publishing
is a trademark of
International Book Market Service Ltd., member of OmniScriptum Publishing Group
17 Meldrum Street, Beau Bassin 71504, Mauritius

Printed at: see last page
ISBN: 978-613-9-45090-9

FABRICATION OF FLEXIBLE SCREEN-PRINTED DIELECTROPHORETIC DEVICES

by Marius Andrei OLARIU

2019

Preface

A brief presentation of the technological steps which have to be followed in order to develop a flexible screen-printed dielectrophoretic device is provided within the following four chapters of the book. The main objective of the lecture is to familiarize the master students with the basic theoretical and practical notions regarding the screen printing, ink or paste formulation, flexible substrate manufacturing, dielectrophoresis as well as common factors affecting the printing processes in general. The content is more a collection of technical best practices which the author acquired during his research and development work with screen printing technology and after synthesizing and reviewing information within a plethora of specialized studies.

It is my hope that this book will be a useful information tool for graduate students aiming to commence working in the field of flexible electronics.

The author is expressing his gratitude to Arcire Alexandru who provided un-measurable technological help by fully dedicating in the practical work described herein.

The author

Table of Contents

Chapter I – Printing technologies for flexible electronics

I.1   Introduction   pp.7
I.2   Overview of printing technologies   pp. 7
I.3   Screen-printing technology   pp. 14
        References

Chapter II – Printing ink or paste formulation

II.1   Metal based inks   pp. 23
II.2   Carbon based inks   pp. 25
II.3   Polymer-based inks   pp. 27
II.4   Encapsulation inks   pp. 29
        References

Chapter III – Technology for flexible polymeric substrate

III.1   Sol-gel method   pp. 33
III.2   Conventional polymers employed as matrix for   pp. 38
        nanocomposite development
III.3   Semi-crystalline thermoplastic polymers   pp. 40
III.4   Non-crystalline thermoplastic polymers   pp. 45
        References

Chapter IV - Dielectrophoretic manipulation of particles and living cells

IV.1.   Theory of Dielectroforesis   pp. 57
IV.2.   Examples of dielectrophoresis set-ups for nanoparticles pp. 65
        and living cells manipulation
IV.3.   Practical utilization of dielectrophoretic nanoparticles and   pp. 67
        living cells manipulation

IV.4.   Hands-on design and fabrication of dielectrophoretic   pp. 72
        screen printed devices
        References

List of Figures

Figure 1.1 – Schematic representation of ink-jet printing

Figure 1.2 – Schematic representation of slot-die coating

Figure 1.3 – Schematic representation of gravure printing

Figure 1.4 – Schematic representation of offset printing

Figure 1.5 – Schematic representation of nano-imprinting

Figure 1.6 – Schematic representation of aerosol jet printing

Figure 1.7 – Schematic representation of screen printing

Figure 1.8 - Screen mask with unbias angle mesh

Figure 1.9 - Screen mask with bias angle mesh

Figure 1.10 - Screen mask with bias angle mesh

Figure 1.11 – Example of geometric shapes of blades

Figure 3.1 – Sol-gel process

Figure 3.2 – Comparison between contact angle of a the hydrophobic and hydrophilic surfaces

Figure 3.3. – Glass transition temperature for a series of classical polymers employed as flexible substrate

Figure 4.1 – Schematic representation of electrophoresis (EC)

Figure 4.2 – Interfacial or Maxwell-Wagner polarization in case of living cells

Figure 4.3 – Dipolar or orientation polarization of water molecule

Figure 4.4 – Graphical representation of DEP a) positive and b) negative

Figure 4.5 – Graphical representation of Re[K(ω)] vs frequency for an arbitrary target particle

Figure 4.6 – Schematization of DEP set-up

Figure 4.7 – Common biological cells dimensions

Figure 4.8 – Basic IDE structure

Figure 4.9 – Numerical simulation of (a) real permittivity and (b) loss factor evolution against the frequency

Figure 4.10 – Distribution of the electric field at the level of IDE with rectangular castellated microelectrodes

Figure 4.11 – Vector representation of the electric field at the level of IDE with rectangular castellated microelectrodes

Figure 4.12 – Distribution of the electric field at the level of IDE with saw-shaped fingers

Figure 4.13 – Vector representation of the electric field at the level of IDE with saw-shaped microelectrodes

Figure 4.14 – Numerical evaluation of mechanical behaviour according to von Mises yield criterion – IDE with castellated fingers

Figure 4.15 – Numerical evaluation of mechanical behaviour according to von Mises yield criterion – classical IDE configuration

Figure 4.16 – Image of the screen mask: a) the side of the mask in contact with the printing substrate, b) the side of the mask in contact with the squeegee

Figure 4.17 – Manual screen printer

Figure 4.18 (a) – Silver based IDEs screen printed

Figure 4.18 (b) – Detailed picture of silver based IDEs screen printed

Figure 4.19 – Basic dielectrophoretic set-up

Figure 4.20 – DEP experimental laboratory bench

Chapter I – Printing technologies for flexible electronics

I.1. Introduction

Flexible electronics also known as organic electronics is defining a category of electronic devices developed on a bendable substrate material (e.g. paper, polymers) with the aim of addressing the current societal needs in terms of applications (e.g. increasing the energy production from renewable sources, improving energy efficiency, providing new diagnosis tolls for health system), target which can be reached only by diminishing the production cost of the electronic and electrical devices. As a general notion, a printed electronic device is encompassing *the substrate, electrical connections, passive and/or active circuit element and the encapsulation.* The flexible substrate is mainly the most important component of this category of devices. Different from classical electronic devices, the active components of a flexible electronic device are not mounted but should be developed directly as part of a single production process which is presuming a series of sequential manufacturing stages. For developing all these components many printing technologies have been reported within the scientific literature and Khan realized a very good classification of the printing technologies as *contact and non-contact printing methods.* The category of non-contact technologies is encompassing *screen printing, ink-jet printing and slot-die coating* while the contact printing category is including *gravure, offset, flexographic, micro-contact, nano-imprinting (embossing) and dry transfer printing*. This classification is realized based on the technological approach selected for transferring an image as pattern at the level of a substrate. For mass production, *roll-to-roll printing* has emerged as the most convenient printing technologies both from costs and quality viewpoint.

The herein chapter is presenting an overview of the both non-contact and contact technologies employed at the time being in flexible electronics development while afterwards the screen-printing method is detailed.

I.2. Overview of printing technologies

Screen printing is mostly considered the simplest method for developing electronic devices. The technology is presuming the transfer of a certain material at the level of a substrate through a screen mask containing openings under the shape of the pattern to be transfer. The printed pattern is realized by solidification of the ink/paste which is transferred at the surface of a substrate through the opening of the screen mask.

Ink-jet printing is a process computerized assuming the development of a certain pattern on the surface of a substrate through controlled disposal of ink droplets in accordance with the digital image/shape elaborated by the human operator. The method was firstly used in publishing industry and was afterwards adopted by office printing devices and flexible electronics industry.

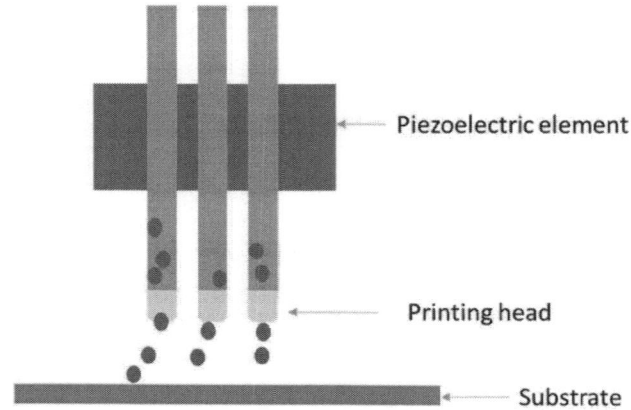

Figure 1.1 – Schematic representation of ink-jet printing

Slot-die coating is the simplest technology adapted for roll-to-roll and roll-to-plate printing. The process is assuming the utilization of an ink dispenser provided with one or more slots through which the liquid material is transferred at the level of the target substrate under the shape of lines. The dimensions of the lines patterned is varied by the number of slots.

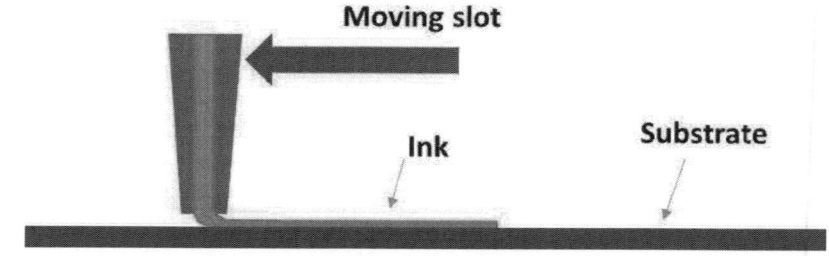

Figure 1.2 – Schematic representation of slot-die coating

The process is suitable for mass production of large scale electronics (e.g. photovoltaics) as of the good uniformity of the coating. However, the resolution of the printing is difficult of being improved. Another advantage of the technique is provided by the fact that the thickness of the pattern can be varied from the level of nanometres to micrometres, and besides, the ink losses are very low.

Gravure printing is a relative simple printing process which is allowing the transfer of patterns with very good resolution at relative low speeds. The most important components in gravure printing are the impression and gravure cylinders. At the surface of the gravure cylinder through chemical, electrochemical, laser and/or mechanical technologies the shape of the pattern to be transfer is developed. The gravure cylinder is in general developed of soft materials as copper and is covered with a deposit of hard materials (chrome) for avoiding eventual damages of the shape to be transferred. The advantages of gravure printing are provided by the possibilities of employing inks of various viscosities. Besides, the quantity of the transferred ink is big, and the printing speed which is very high and the very good resolution of the printed pattern. On the other hand, the technology is presenting a series of drawbacks as the cost for engraving the gravure cylinder are very high, the time needed for pre-arranging the process equipment is very long. The technology is suitable for high volume printing and not suitable for fragile substrates as of the pressing force of the gravure cylinder.

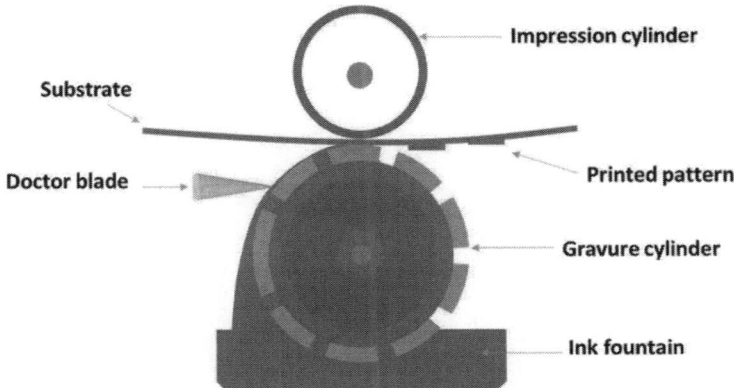

Figure 1.3 – Schematic representation of gravure printing

Offset printing was established in order to avoid the main drawback of the gravure printing, a new technology was established. The method is assuming the utilization of a gravure cylinder which is not directly transferring the ink to the substrate but at the level of a blanket cylinder which is in direct contact with support material.

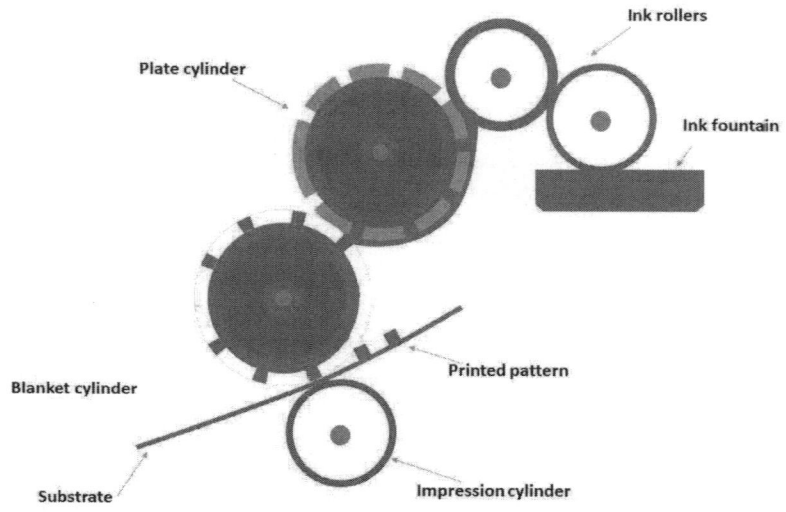

Figure 1.4 – Schematic representation of offset printing

As the gravure printing as well, the offset printing can be used for printing on flat substrate or on a substrate transported by a cylinder carrier. The difficulty of the technology is provided by the complexity of the mechanisms of the equipment, but the advantages in comparison to other printing processes are the very good resolution of the printed patterns and high printing speed. The drawback of the method is the limited life cycle of the printing plate fact which is increasing the production costs.

In the case of **flexographic printing,** the anilox cylinder represents the most important component of the equipment. The anilox cylinder is generally realized of stainless steel or aluminium and covered by a ceramic layer engraved with cells (dimples). The ink is firstly filling the cells of the anilox roller which is afterwards transferring it at the level of a printing roller provided with elastic covering and engraved with the image to be patterned. The technology can be employed for both soft and hard substrates. The quality of the pattern transferred is improved as of the fact that the ink is uniformly distributed at the level of the anilox roller as of the dimples, but the elastic covering of the printing roller may be subject to early degradation which may lead to a diminishing of the operation time of an equipment.

In the case of **micro-contact or micro-transfer printing,** the elastomeric stamp which is the tool for transferring the ink at the level of a certain support is the key component. Generally, the stamp is realized of polydimethylsiloxane (PDMS) which in the liquid estate is spread at the level of a Si wafer. After solidification through photolithographic techniques the mold is realized. The quality of the photolithographic mask is significantly influencing the quality of the mold and the one of the pattern.

The technology is not difficult being suitable for laboratory practices and has the advantage of allowing development of replica molds. However, special attention should be paid as during the transfer of the ink deformation of the stamp may occur and besides, the solvents in the ink may also affect the PDMS mold.

A derived printing process from micro-contact printing is the **laser micro-transfer printing**. The set-up of laser micro-transfer printing is the same as the one of the micro-contact printing the difference is the one that the transfer of the ink is not realized mechanically through pressing but through a pulsed laser beam directed toward the interface between the stamp and the microstructure to allow the ink to adhere to the substrate.

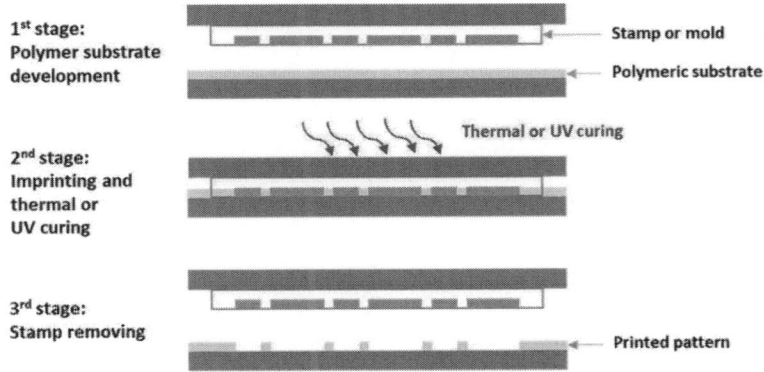

Figure 1.5 – Schematic representation of nano-imprinting

Nano-imprinting (embossing) was proposed for the first time in 1995 as a process of transferring various patterns of micro and nano- dimension at the level of a target substrate with the help of a stamp with imprinted pattern of

11

nanoscale dimension based on direct mechanical deformation process. The stamp (also known as mold) is the key element of the process being realized by through photolithographic techniques. The curing of the printed shape can be realized through thermal or UV treatment. The use of photolithography in developing the stamp has the advantage of improving the resolution, dimensions and quality of the printed pattern but also is complicating the process.

The thermal expansion coefficient of the material employed for mold fabrication and the material used for substrate development should exhibit close values in order to avoid defects at the level of the printed pattern. Moreover, the hardness of the material used in fabricating the stamp is critical. Both metals and polymers have been demonstrated to be suitable for mold fabrication. However, the polymer based stamps have the advantage of less expensive production costs but have the disadvantage of been susceptive to rapid deterioration and contamination. Within the last years the technology was successfully adapted for developing roll-to-plate and roll-to-roll technology based on flexible stamps. The pressure applied for transferring the pattern is critical during the manufacturing process. Nickel based stamps have been used for developing cylindrical molds for roll-to-roll and/or roll-to-plate nanoimprinting.

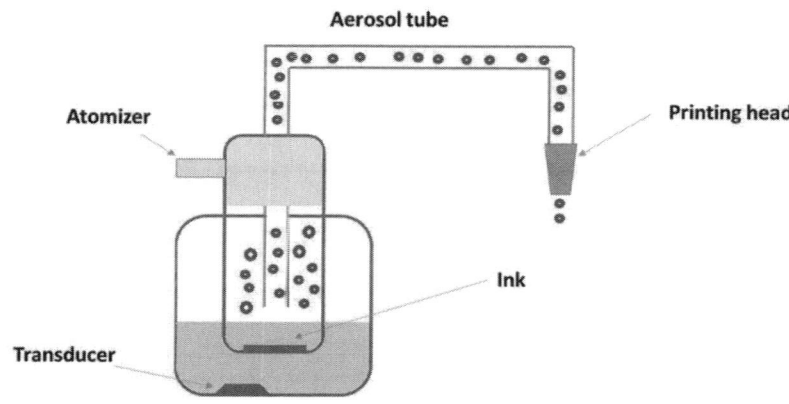

Figure 1.6 – Schematic representation of aerosol jet printing

Dry transfer printing is suitable for transferring nanoscale patterns and is presuming the „sandwiching" of the printable layer between two other substrates. Practically, the printable pattern is grown on the top transfer layer.

Afterwards, a lamination process is realized. As of the adhesion which should be larger between the printable layer and the target substrate during the delamination process, the printable layer will dettach from the top substrate and will adhere at the target substrate. The technology is often employed for developing multilayer active printable layers and for developing solid printable layers as graphene transfer from the substrate it was grown on the targeted substrate.

Aerosol jet printing is a relatively new printing process presuming the use of an atomizer which generates mist with the help of an inert gas (e.g. N_2).

The smallest droplets are directed as of the inert gas through a pipeline through the printing head. Before reaching the printing nozzle, the stream of droplets is increased in density and the inert gas removed. The pattern is realized by displacing the substrate which is fixed on a table which allows the precise motion control. Besides the advantage of contactless and maskless printing capabilities, aerosol jet printing is allowing the use of inks of various viscosities which may vary in a broad range from 1 to 1000 cP. **Xerographic printing** which is basically the classical electrophotographic process developed by Xerox decades ago employing liquid toner instead of solid one was adopted as technology for flexible electronics development. The technology was succesfully used for printed memory development.

Plasma jet printing has at key component a high voltage electrode and a gravure cylinder as a second electrode. The stream of dimples is directed towards the target surface with a carrier gas through multiple needles. The process is realized at 40°C and consequently a large variety of materials (from paper to polymers and metals) can be used for target substrate manufacturing. Moreover, the technology is suitable for 3D printing as well.

I.3. Screen-printing technology

The **screen-printing** represents a process of transferring a material in a liquid estate under the pressure of an external force through a dedicated printing mask at the level of a solid material used as support. Following the transfer, the liquid, eventually in a controlled manner (e.g. thermal treatments) is adhering at the surface of the solid material and is solidifying as permanent or semi-permanent pattern.

The **screen mask** used for developing the pattern is in general a rectangular frame with a mesh attached on it. The frame is generally realized of wood or metallic material. However, wood is not so used in electronics manufacturing as it is subject to an early deterioration. Thus, presently, the frame of the screen mask is realized of metal, especially aluminium as of its low manufacturing cost as well as robustness and lightweight. There are also producers as Kuroda Electric which are employing ceramic material for frame development.

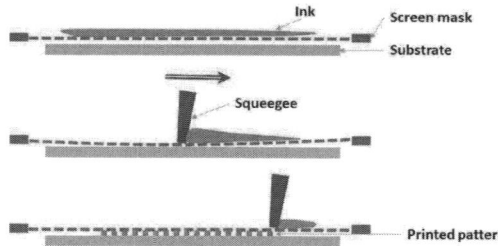

Figure 1.7 – Schematic representation of screen printing

Depending on screen-printing dimensional constraints the dimension of the frame can vary. However, the producers established standards for screen frame dimensions and shapes as well. The open area of a screen mask can be calculated with the formula:

$$Open\ Area = \frac{mesh\ area^2}{(wire\ diameter + mesh\ opening)^2} * 100\%$$

where the mesh area is the surface of the mesh attached to the frame (inner area of the frame), wire diameter is the gauge of a wire and mesh opening is the distance between two warp or weft wires.

The mesh screen is critical in developing the screen mask not only from dimensional perspective, but also from the viewpoint of the materials used for developing the mesh as well as from attaching technology of the mesh to the frame. The mesh used for developing screen mask for electronic industry are generally fabricated of stainless steel, tensioned stainless-steel, polyester, polyamide-nylon, polyarylate.

When selecting the mesh, the application or final product should be considered in order to ensure the fact that the ink used is transferred trough mesh's openings in order to not obstruct (permanently or temporary) the openings. The thickness of final pattern or deposit thickness is a parameter depending on mesh thickness as well as emulsion thickness.

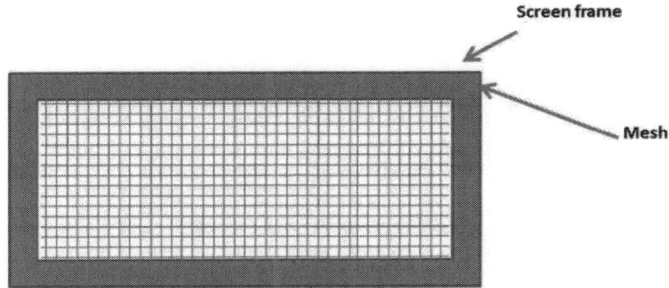

Figure 1.8 - Screen mask with unbias angle mesh

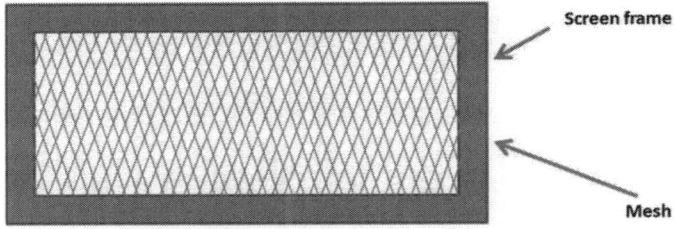

Figure 1.9 - Screen mask with bias angle mesh

The most important parameters which should be considered when selecting a mesh are:

1. Mesh opening or aperture size - the distance between two warp or weft wires.

2. Mesh count – the number of mesh openings per square inch or millimetre.

As a general rule, in screen printing process, in order to ensure correct transfer of ink through the screen-stencil, the mesh diameter should be at least three times larger than the size of particles in the printing ink.

3. Overall open area of the screen mesh – is the sum of all openings of the mesh attached to the screen frame.

4. Mesh thickness – is the thickness of a mesh after warping or weaving.

5. Mesh bias angle – is the angular position of the mesh in respect to the screen frame.

6. Emulsion thickness over mesh (EOM) – is the final thickness of the mesh after obstructing specific openings in order to develop the needed stencil-mesh or pattern and is influencing the thickness of the target deposit or pattern.

In respect to modality of attaching the mesh to the screen frame, two types of screen masks can be distinguished:

1. Direct mesh to mask system, is the most common attaching system, within which the mesh is directly fixed on the screen frame.

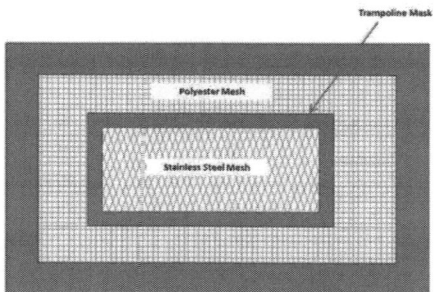

Figure 1.10 - Screen mask with bias angle mesh

2. Trampoline mask, is a more dedicated attaching system within which the useful mesh is connected to the mask frame through an intermediary mesh of more flexible wire (e.g. polyester). This attaching system is practically increasing the mask lifetime but during the screen printing process is attenuating the stress caused by squeegee movement.

Case a thicker thickness of the target deposit is needed; one can choose a 3D mesh which is a mesh obtained through modification of warping or weaving process. On the other hand, for developing target deposit of lower thickness, a **calender mask** is suitable, meaning a mask with a flattened mesh. One-side calender mask are also available on the market. During the flattening process, the opening area is affected as well.

Depending on the pattern to be transferred shape, the mesh should be provided with unobstructed and obstructed openings. The obstructed openings are realized with impermeable solidifying solutions. The characteristic of mesh employed is critical in mask development process and is influencing the printing as a whole.

The pattern to be transferred is known under the name of **stencil design**. The stencil design is fabricated on the mesh with the help of various emulsions. The emulsion and the mesh openings are determinant for diminishing the width to be printed as well as overall contour quality.

For developing the stencil design, the screen mesh should be completely coated with photo-emulsion. After coating, the mesh should be exposed to light. As the coating is photosensitive, between the UV source and the screen mask a positive drawing on a transparent films should be inserted. The UV will allow the area coated with emulsion to harden while the covered emulsion will be removed allowing the development of the stencil design through which ink will be transferred at the level of the substrate in a further stage.

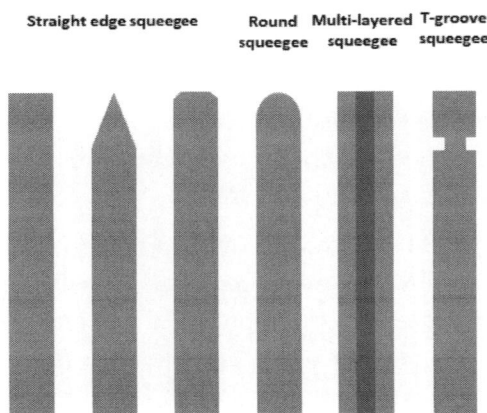

Figure 1.11 – Example of geometric shapes of blades

Another element which is crucial in screen printing is the squeegee which is used to transfer through the screen mask the ink for pattern development. The most important parameter in selecting the right squeegee is the rigidity of the blade which is measured by durometer which is expressing the hardness of the blade. The higher the durometer will be, the harder the blade will be as well. The durometer of the blade is provided by the materials employed for its fabrication. In the majority of case, the material used for blade fabrication is natural rubber, polyurethane or neoprene. The polyurethane squeegees are mainly used for semi-automatic or automatic screen-printing.

On the other hand, the shape of the blade may differ. The square or plain (with straight edge) blades are most often used as they are proper for regular inks.

The rounded blades are used when a larger volume of ink is to be transferred. Finally, the v-shaped or angled blades are common for cylindrical screen printing. Multi-layered blades are also very wide-spread. The blade of this squeegee is manufactured of two or three layers of materials of different rigidities sandwiched. Generally, the middle layer is the harder and is practically ensuring that the squeegee is not bending during utilisation. On the other hand, from practical viewpoint, it should be mentioned that the width of the squeegee should be at least one centimetre on each side than the printing area, but should not get to close to the screen frame as it can affect the tension of the screen mesh.

In respect to movement of the squeegee on the screen mesh during the screen printing process, the angular position of the squeegee is determinant. A very good transfer of the ink at the substrate level is obtain case the angle between the squeegee and the screen mesh is close to 45°. A higher angle will not allow the transfer of a proper volume of ink at the level of the substrate, while a narrow angle will practically increase the quantity of ink to be transferred, obstructing eventually the mesh's openings.

Another crucial parameter in screen-printing process is the snap-off distance which is the distance between the screen mask and the substrate before the printing and is an important factor in improving the shear rate factor. The snap-off distance is directly dependent on the mesh tension, movement of squeegee, type of material used for screen mask frame, thickness of the substrate and the material used for substrate development.

Besides, the characteristic parameters of the final product obtained through screen-printing are strongly dependent on ink's characteristics (e.g. viscosity), type of squeegee employed for transferring the ink, substrate's surface. The ink employed in the majority of cases for screen-printing should exhibit a high viscosity in order to avoid its leaking through the mesh. In many cases it is also called paste. The paste which will be transferred at the surface of the substrate should be a duplication of the stencil design.

Technical specifications of an ink are or may include:

1. viscosity which can be provided in cps (centipoises) or mPa*s (milipascal second).

2. surface tension in mN/m.

3. total percentage of solids in ink provided as percentage.

4. volume resistivity in Ωcm and surface resistivity in Ω.

5. maximum time and conditions for being deposited.

6. grind gauge (also known as grindometer or Hegman gauge), meaning the fineness of the particles dispersed in the ink or paste, and is provided in micrometers (µm).

Hegman units	Microns
0	101.6
1	88.9
2	76.2
3	63.5
4	50.8
5	38.1
6	25.4
7	12.7
8	0

7. drying time in seconds at a certain temperature.

8. flash point, meaning the lowest temperature at which the vapours of the material will ignite.

9. shear rate is the velocity of the squeegee divided by the distance between the squeegee and the substrate.

Additional to all these, information on the paste after post-printing and post-treatment might be provided by ink manufacturers.

19

In the screen-printing process, the paste is dosed with the help of a flood bar or distributor, but pipettes or syringe can be used as well. Case the viscosity is too small (200 mPa*s), the ink will directly leak from the mesh. Thus, the mesh opening and the ink viscosity should be correlated. Moreover, the dimensions of the particles dispersed in the paste should be carefully selected,

The advantages of screen-printing derived from the fact that the technology is the most developed conventional printing method, and is allowing the rapid patterning of various materials, being very versatile. Moreover, pattern geometries can be varied and improved resolutions of the pattern can be obtained due to plethora of commercial available meshes currently on the market. However, the most important advantage of screen printing is that the fabrication costs are much lower in comparison with other printing methods.

On the other hand, a series of drawbacks have been encountered in practice when employing screen printing, and most important issues are related to the fact that the ink requires a certain time for drying. Various drying techniques or post-process techniques can be used to facilitate ink's solidification at the level of the substrate. Moreover, for avoiding eventual leakages through the mesh, the ink should have a high viscosity while a short drying time may lead to potential dryings on the mesh.

References

[1] Saleem Khan, Leandro Lorenzelli, Technologies for Printing Sensors and Electronics Over Large Flexible Substrates: A Review, IEEE SENSORS JOURNAL, VOL. 15, NO. 6, JUNE 2015.

[2] Zheng Cui et. al, Printed Electronics: Materials, Technologies and Applications, Wiley, September 2016, ISBN: 978-1-118-92092-3.

[3] Giovanni Nisato, Donald Lupo, Simone Ganz, Organic and Printed Electronics: Fundamentals and Applications, 1st Edition, Pan Stanford, March 2016, ISBN 9789814669740.

[4] Ostfeld, A. E. et al. Screen printed passive components for flexible power electronics. Sci. Rep. 5, 15959; doi: 10.1038/srep15959 (2015).

[5] Beatrice Medina Rodriguez, Inkjet and Screen Printing for Electronics Applications, Ph.D. Thesis, September 2016, Universitat de Barcelona.

[6] Ahmed Tausif Aijazi, Printing Functional Electronic Circuits and Components, Ph.D. Thesis, December 2014, Western Michigan University.

[8] Joseph Chang et. al., Challenges of printed electronics on flexible substrates, Circuits and Systems (MWSCAS), 2012 IEEE 55th International Midwest Symposium on, DOI: 10.1109/MWSCAS.2012.6292087.

[9] T. Kaufmann and B. J. Ravoo, "Stamps, inks and substrates: Polymers in microcontact printing," Polym. Chem. , vol. 1, no. 4, pp. 371–387, 2010.

[10] A. Nathan et al., "Flexible electronics: The next ubiquitous platform," Proc. IEEE, vol. 100, pp. 1486–1517, May 2012.

[11] R. S. Dahiya, "Towards flexible and conformable electronics," in Proc. 10th Int. Conf. Ph.D. Res. Microelectron. Electron. (PRIME), Grenoble, France, 2014, pp. 1–2.

[12] William S. Wong, Alberto Salleo, Flexible Electronics: Materials and Applications, Springer 2010.

[13] Zhigang Wu et. al, Opportunities and Challenges in Flexible and Stretchable Electronics: A Panel Discussion at ISFSE2016, Micromachines 2017, 8, 129; doi:10.3390/mi8040129

[14] Valerio Zardetto, Thomas M. Brown, Andrea Reale, Aldo Di Carlo, Substrates for Flexible Electronics: A Practical Investigation on the Electrical, Film Flexibility, Optical, Temperature, and Solvent Resistance Properties, JOURNAL OF POLYMER SCIENCE PART B: POLYMER PHYSICS 2011, 49, 638–648

[15] Jing Wu and Min Gu, Microfluidic sensing: state of the art fabrication and detection techniques, Journal of Biomedical Optics 16(8), 080901 (August 2011).

[16] K. D. Harris, A. L. Elias, H.-J. Chung, Flexible electronics under strain: a review of mechanical characterization and durability enhancement strategies, J Mater Sci (2016) 51:2771–2805, DOI 10.1007/s10853-015-9643-3.

[17] Hauger TC, Al-Rafia SMI, Buriak JM (2013) Rolling silver nanowire electrodes: simultaneously addressing adhesion, roughness, and conductivity. ACS Appl Mater Interfaces 5:12663–12671

[18] Hyejin Park et al 2012, Fully roll-to-roll gravure printed rectenna on plastic foils for wireless power transmission at 13.56 MHz, Nanotechnology 23 344006, DOI:https://doi.org/10.1088/0957-4484/23/34/344006

Chapter II – Printing ink or paste formulation

One of the most challenging tasks in printed electronics is the identification of the most convenient equilibrium between the technical characteristics of the ink or paste and the final printed product in order to ensure reliable functionality of the product as well as facile operability of the overall printing process.

Screen printing inks for printed electronics are available commercially but in order to accomplish the desired functions they may be used as purchased or should be tailored in accordance with their aim. The inks for screen printing are categorized as *metal based inks, carbon based inks, polymer-based inks and encapsulation inks.*

II.1. Metal based inks

Metal based inks are the mostly employed inks mainly as of the fact that do exhibit very good electrical conductivity. *Silver, copper, gold, platinum, aluminium, brass, nickel, chrome, iron, titanium* are among the most usual materials used in formulating metal based inks. In any printing process, the ink or paste should exhibit a series of technical characteristics which for ensuring *good compatibility* and *adhesion of the printed pattern*.

According to Nir, in order to satisfy technical expectations, the conductive inks should demonstrate a curing or annealing temperature below 150 °C (preferably below 120 °C) to not overstress thermally the polymeric based flexible substrate. Besides, in order to improve reproducibility of the printing process, the metallic particles *should not aggregate*. For this purpose, within the ink a *stabilizing agent* can be added.

Besides all these, a good metallic based ink which may ensure high level printing quality of the transferred pattern should demonstrate *good viscosity, surface tension and wettability*. For ensuring good electrical conductivity, the particles employed in paste formulation *should oxidize as slow as possible*. This is the reason why the majority of metal particles used are of noble materials. The non-noble materials are oxidizing within less than a second fact which is critically affecting the conductivity. Of all noble materials, silver is the most common as of its good resistance to oxidation as well as very good electrical conductivity. However, the electrical conductivity of a silver based ink is three or four time lower than the one of the pure metal. According to Kela, the conductivity of Ag based inks is 4-5 times lower than pure metal and price two to three times of bulk Ag. Besides, another disadvantage of the silver based ink is the high production price which is also varying a lot in accordance with the market trends.

Many attempts have been also reported lately for replacing silver with copper. Copper is a material is more abundant at worldwide level and is much cheaper but it has a higher affinity to oxygen meaning that it will oxidize faster. Kee splitted the copper-based conductive inks in three categories: *traditional micro-sized flakes and powders type, nanoparticle type and precursor type*. In case of filler-type conductive ink, sintering temperature is about 80% of the melting point of the filler material in the inks. For this reason, the materials employed for substrates manufacturing is limited to alumina, aluminium nitride and silicon nitride. On the other hand, in case of filler-type ink, the temperature for sintering the nanoparticles is much lower, up to <300 °C.

The most appropriate ink for polymer based substrates is the one based on metal nanoparticles as the sintering temperature is comprised between 100 and 300°C. Overall, in order to avoid a rapid oxidation of the copper nanoparticles the sintering should be realized within a hydrogen or nitrogen environment or intense-pulsed light (IPL) sintering should be applied.

According to literature proper inks based on nanoparticles should be realized and cured at low temperatures.

In case of cobalt for example, a convenient content within the ink for screen printing should be around 70 wt. % in order to allow good line width and uniformity of the printed pattern and the most homogenous distribution of the particles in the printed patterns.

The preparation of the metallic based inks can be realized through two technical approaches: (a) **top-down methodology** meaning physical methods supposing the reduction of crystal size or (b) **bottom-up based on classical chemistry.**

In case of top-down methods, larger quantities of nanocrystals can be realized, while in the bottom-up approach the nanocrystals can be synthesized with controlled particle size. However, bottom-up ink preparation process is a much more difficult process assuming many production stages, and thus, the wet chemistry methodology is much convenient of being employed as nanoparticles should be only dispersed in suspension but also in economic terms.

In 2016, Black proposed a Reactive Organometallic (ROM) ink formulation for the sinter-free printing of high conductivity crystalline silver films. The process employs Atomic Layer Deposition (ALD) to develop new ink formulation.

0.5 M:0.5 M organometallic-alcohol ROM ink formulation developed allowed the fabrication of silver based films with an electrical resistivity down to 4.1×10^{-8} Ωm, meaning approximatively 40% less that the ones of bulk silver.

II.2. Carbon based inks

The carbon based inks are a viable alternative to metallic based inks as the manufacturing costs of carbon based fillers are decreasing lately. The classical composition of a carbon based ink is assuming utilization of four elements: *solvent, filler, dispersant and binder.*

Single walled (SW-) and multi walled (MW-) carbon nanotubes (CNTs) have been both employed for developing carbon based pastes.

A CNT based ink is assuming the fulfilment of classical technical characteristics of an ink as low viscosity and high share rate. However, the stability of a CNT based suspension is difficult of being realized as of the fact that CNTs aggregate in the presence of Van der Waals forces. In order to stabilize a CNT based ink, Biswas proposed the utilization of sodium dodecyl sulphate (SDS) as dispersant and polyvinyl pyrrolidone (PVP). Menon reported the preparation of MWCNT based ink which supposed firstly the addition of SDS into 5 ml ethanol while afterwards 7.5 wt % of MWCNTs were added in the mixture which was stirred for 24h. For ensuring proper debundling of MWCNTs and long-time stabilization of the solution the binder (PVP) was added. Case the binder is correctly chosen the viscosity and electrical properties of the ink are not affected. It is obvious that the quantity of binder should be tailored in accordance. Menon reported that after screen-printing, the conductive patterns on flexible substrates exhibited very good adhesion characteristics and a resistance in the range of 0.5–13 Ω sq^{-1} with minimum surface roughness (<65nm).

Tortorich employed pristine single-walled carbon nanotubes (SWCNT) and sodium n-dodecyl sulfate (SDS), an anionic surfactant, for developing a carbon based conductive ink. The aqueous solution contained 0.8 mg/mL of SWCNTs and 3 mg/mL of SDS and was ultrasonicated for 30 minutes. After sonication, the carbon nanotube ink was centrifuged at 12,000 rpm for 5 minutes. According to the author, the dispersant concentration is extremely important in preparing carbon nanotube based ink as even if higher dispersant concentration may result in a well-dispersed carbon nanotube ink, after printing the patterns resistance may be increased.

On the other side, a low quantity of dispersant, will not allow CNTs to be dispersed properly. Consequently, the CNTs concentration should be tailored in accordance with the dispersant used. Besides, the SDS concentration should be correctly chosen as it is directly influencing the surface tension of water-based CNT ink.

Zhang correctly summarized the main advantages of carbon based inks as: "easy renewal, easy modification and reproducibility". It was also highlighted the fact that the electrochemical and electroanalytical properties of the printing are affected not only by the binder but also by the carbon materials. Besides, the binder is also influencing the printed pattern mechanical stability, lifetime, as well as electrochemical inactivity.

Phillips proposed a conductive ink based on graphite and carbon black (CB). The two carbon based particles were added gradually, first the carbon black and afterwards the graphite, and homogenization was realized by hand. Phillips demonstrated that the simultaneous utilization of carbon black and graphite is leading to increased conductivity of the final ink as of the fact that the CB is bridging the graphite agglomerations.

The highest conductivity of the ink was noticed when the carbon quantity reached 29.4% while the ratio of graphite to CB 2.6 to 1. Lower carbon content is requesting for lower ratio of graphite to CB. However, an increase of the CB content leaded to a "higher rest viscosity and greater shear thinning and a lower high shear viscosity".

Graphene, as the first 2D material, has been also employed in the preparation of carbon based inks. Graphene based ink can be developed in the presence or absence of binders. The binder-free methods are preferred as of the fact that the thermal annealing is realized at lower temperature even if the method employing binders is allowing the formulation of inks of higher conductivities.

Huang reported the development of a binder-free conductive ink method and claimed that the new developed ink is suitable for flexible substrates as paper and textiles. On the other hand, graphene oxides (GO or GOx) was not neglected for preparation of conductive inks even if the material exhibits lower conductivity in comparison to graphene. The GO is preferred as of its hydrophilicity, very good solubility in water to the fact that the derived solutions containing GO are much more stable.

II.3. Polymer-based inks

Within the last years conjugated polymers have attracted tremendous interest as of their structure presenting alternating single and double bonds in the polymer chain which is allowing hoping conduction to occur case charges are submitted to electrical fields. In order to transform a conjugated polymer which is basically insulator from electrical viewpoint into a conducting polymer (CP) or semiconducting polymers (SCP) doping is a viable process employed as alternative. Through reductive or oxidative process an "n-" or "p-"doped polymer exhibiting electrical conductivity ranging from 10^{-8} to 10^3 S/cm can be processed. The electrical conductivity of the polymers is practically critical when developing screen-printable devices, and under these circumstances, efforts have been driven towards increasing it. Through doping, within the conjugated polymers, **polarons** may be formed as of the "interaction" between the electrons and/or holes and their self-induced polarization with the polar dielectric. The displacement of the charges along the main chemical chain is fostering the generation of an electric current. Within this category of materials polyacetylene (PA) was firstly reported about 30 years ago by Hideki Shirakawa, Alan Heeger, and Alan MacDiarmid. Afterwards, polymers as polyaniline (PANI), polypirrole (PY) and polyparavinylene (PPV) have emerged. The synthesis of the conducting polymers starting from conjugated polymers can be realized though chemical or electrical polymerization. The chemical polymerization has the advantage of being suitable for mass-production as of the fact that the polymerization is occurring almost instantly and the price of the materials employed are much lower. However, the electrochemical polymerization has the advantage that the resulted conductive polymer is basically exhibiting higher electrical conductivities, but its employment is much more difficult to be realized.

As of the low electrical conductivity in comparison to classical metals, the conducting polymers were employed mainly in fabrication of sensing and biosensing devices. The importance of utilizing conducting polymer in development of biosensing devices is nowadays widely recognized. Screen-printed polyaniline electrodes were developed by Gosselin at its team with the aim of fabricating a potentiometric sensor for real-time monitoring of a LAMP (Loop-mediated isothermal amplification) reaction.

Poly(3,4-ethylenedioxythiophene) doped with polystyrene sulfonate acid or shortly PEDOT:PSS is an intrinsically conductive polymer intrinsically used presently as an alternative to classical ITO (Indium Tin Oxide). PEDOT:PSS is mostly known and used as of its outstanding properties as optical transparency, high electrical conductivity, electrochemical stability, viscosity, low surface roughness, good flexibility, as well as reasonable processing cost. Many studies are reporting continuous interest towards improving the electrical conductivity of PEDOT:PSS through addition of polar solvents or through utilization of conductive fillers as CNTs. In spite of the fact that PEDOT:PSS is exhibiting a lower electrical conductivity in comparison to ITO which unfortunately cannot be processed as ink. PEDOT:PSS is much more flexible than ITO and is relatively easy to be processed as ink. PEDOT:PSS may be used as commercial aqueous solution or may be transformed into hydrogel-like ink as reported by Nakashima. The reason of converting PEDOT:PSS into hydrogel is to avoid leakages during the printing process.

PEDOT:PSS based ink was formulated by Perinka to develop printed microelectrodes with thickness ranging from 80 to 750 nm and width of 40um on substrates of PET or PEN.

On the other hand, Sergeev presented the formulation of an aqueous solution based on PEDOT:PSS containing dispersed carbon (soot) particles which exhibited solubility in water, and a obility in the solid layer within a range of 2–6 cm^2 V^{-1} s^{-1} as well as good chemical, mechanical and thermal stability.

Basically, according to Lubianez, in order to increase the electrical conductivity of PEDOT:PSS solutions, polyhydroxy compounds as ethylene glycol, and sulfoxides like dimethylsulfoxide can be used but also amides as N-methylpyrrolidone.

Sensing devices for gases were also reported in the literature. Crowley presented a fully printable polyaniline-copper (II) chloride sensor for the detection of hydrogen sulfide gas. The sensing device is composed of screen printed silver interdigitated electrode (IDE) on a flexible PET substrate with inkjet printed layers of polyaniline and copper (II) chloride. Overall, many polymeric-based materials have been prepared in the last years with the aim of improving electrochemical capacities of printed electrodes.

A nano graphene platelets (NGP) in combination with polyaniline ink was prepared by ball milling method by Xu for development of screen-printable thin film supercapacitor as PANI was demonstrated to be responsible for the pseudocapacitive charge generation.

II.4. Encapsulation inks

The encapsulation inks, often referred to as potting inks, are used in printed electronics to ensure proper protection of conducting printing parts from electrical, chemical and mechanical viewpoint in order to increasing the lifetime of the final product. From functional point of view, the encapsulation layer based on polymer should exhibit high dielectric constant, low dielectric losses, high dielectric breakdown strength as well as flexibility to eventual physical deformations and very good thermal stability. In many cases the encapsulation should demonstrated good transparency (e.g photovoltaics) and should be impermeable to oxygen and water. Water vapour transition rate (WVTR) in g/m is the parameter used in characterising the permeability to water of the encapsulation layer and according to Su in the case of organic electronic devices should reach values lower than $10^{-4} - 10^{-6}$ g/m/d. However, in the case of gas sensing devices, the encapsulation should impede the oxygen and/or the water in getting in contact with the sensing element of the sensor, but should not be a barrier for the target gas. Commercially encapsulation and potting inks are available both as UV-curable and thermally curable. However, the printing process of the encapsulation is not suitable for large electronics but rather for small electronics as chips or interdigitated microelectrodes.

References

[1] Menon H., Aiswarya R., Surendran K.P., Screen printable MWCNT inks for printed electronics, RSC Adv., 2017, 7, 44076–44081, DOI: 10.1039/c7ra06260e.

[2] Tortorich R.P., Song E., Choi J.W., Inkjet-Printed Carbon Nanotube Electrodes with Low Sheet Resistance for Electrochemical Sensor Applications, Journal of The Electrochemical Society, 161 (2) B3044-B3048 (2014) 0013-4651/2014/161(2)/B3044/5.

[3] Zhang X., Cui Y., Lv Z. , Li M., Ma S., Cui Z., Kong Q., Carbon nanotubes, Conductive Carbon Black and Graphite Powder Based Paste Electrodes, Int. J. Electrochem. Sci., 6 (2011) 6063 – 6073.

[4] Huang, X. et al. Highly Flexible and Conductive Printed Graphene for Wireless Wearable Communications Applications. Sci. Rep. 5, 18298; doi: 10.1038/srep18298 (2015).

[5] Phillips C., Al-Ahmadi A. , Potts S.J. , Claypole T., and Deganello D., The effect of graphite and carbon black ratios on conductive ink performance, J Mater Sci (2017) 52:9520–9530, DOI 10.1007/s10853-017-1114-6

[6] Tencha A. et. al., Synthesis of Graphene Oxide Inks for Printed Electronics, 10.1109/YSF.2017.8126608, 2017.

[7] Gosselin D. et.al, Screen-printed polyaniline-based electrodes for the real-time monitoring of LAMP reactions, Anal. Chem., • DOI: 10.1021/acs.analchem.7b02394, 2017.

[8] Crowley, K. H., Morrin, A., Shepherd, R. L., in het Panhuis, M., Wallace, G. G., Smyth, M. R. & Killard, A. J.v(2010). Fabrication of polyaniline-based gas sensors using piezoelectric inkjet and screen printing for the detection of hydrogen sulfide. IEEE Sensors Journal, 10 (9), 1419-1426.

[9] Xu Y., et al, Screen-Printable Thin Film Supercapacitor Device Utilizing Graphene/Polyaniline Inks, Adv. Energy Mater. 2013, DOI: 10.1002/aenm.201300184.

[10] Cinti S., Polymeric Materials for Printed-Based Electroanalytical (Bio) Applications, Chemosensors 2017, 5(4), 31; https://doi.org/10.3390/chemosensors5040031.

[11] Park J. et. al, Electrical and thermal properties of PEDOT:PSS films doped with carbon nanotubes, Synthetic Metals 161 (2011) 523–527.

[12] Nakashima H. et. al, Liquid Deposition Patterning of Conducting Polymer Ink onto Hard, dx.doi.org/10.1021/la203356s, Langmuir 2012, 28, 804–811.

[13] Perinka N., et. al, Preparation and characterization of thin conductive polymer films on the base of PEDOT:PSS by ink-jet printing, 10th International Conference on Solid State Chemistry 2013, doi: 10.1016/j.phpro.2013.04.

Chapter III – Technology for flexible polymeric substrate

The herein chapter is providing an understanding of physical, chemical as well as with electrical properties which are directly influencing the capacity of conductive and dielectric inks of adhering at the surface of dielectric flexible substrate.

Before entering into details, it should be mentioned that at the level of a flexible electronic device, the flexible polymeric substrate is the layout for all the other printed electronic devices. Thus, the adhesion of various inks at substrate's surface is strongly determined by the substrate's synthesis methodology as well as printing method employed for electronic device development.

A viable alternative to classical polymeric materials used nowadays for substrate development are the nanocomposite polymers as the technology employed for their manufacturing is allowing operator's to tailor their properties in respect to their application.

As a rule, a nanocomposite is defined as a material containing at least one nanometric phase (also known as dispersed phased). The organic-inorganic composites are presenting both the advantages of organic components (as flexibility, low dielectric constant and good processability), and of inorganic components as well (durability and thermal stability). To date, a plethora of polymers have been employed for composite development due to their various possibilities of being synthesised and processed, improved adhesion when reinforced, high resistance to corrosive agents as well as reduced density.

The utilization of a nanometric phase is practically allowing the particularization of various properties of the composite material depending on the application. Generally, the properties of nanocomposites are dependent of a series of variables as type of polymeric matrix, the dispersion degree of the dispersed phase, dimension and shape of nanoparticles, orientation of dispersed phased onto the polymeric matrix, as well as interactions between the matrix and the dispersed phase. Overall, the interaction between the nanoparticles and the matrix is strongly affecting the nanocomposite's properties.

III.1. Sol-gel method

At the very beginning, the nanocomposites were developed for improving the mechanical properties of the polymeric matrix. However, the scientific literature is demonstrating that electrical, magnetic, optical and chemical properties may be predefined or tailored in the case of a composite. **The sol-gel method** is one of the most widely spread technique for nanocomposites development, but in general, this technology is allowing the development of nanocomposites with weak intrinsic mechanical properties.

The sol-gel process consists of a hydrolysis reaction of an alkoxysilane for producing hydroxyl groups followed by condensation of hydrolysis products. With other words, the technique is assuming the dispersion of nanoparticles within a liquid while afterwards it would be brought back in a solid state in a controlled manner (generally, thermally or directly through evaporation). The properties of the resulted hybrid product are influenced by the dimension of the particles as well as interaction between the solid and liquid phase. For obtaining organic-inorganic homogenous films, the compatibility between the two components is critical (e.g. introduction of covalent bonds between the organic polymer and inorganic compound).

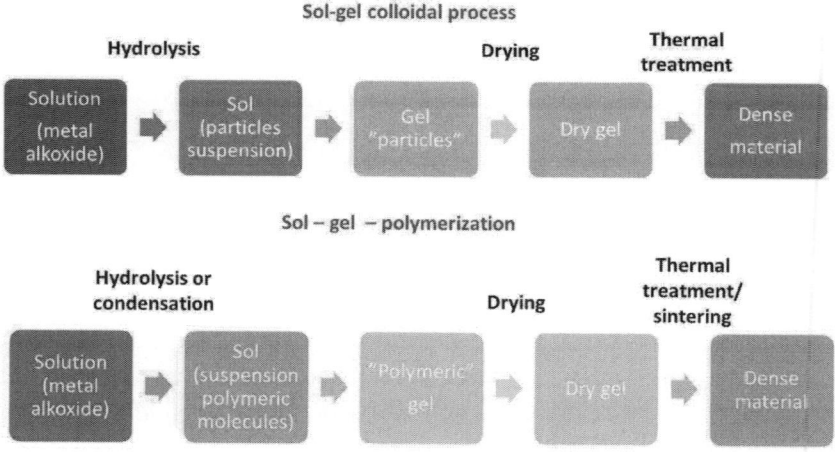

Figure 3.1 – Sol-gel process

A "**sol**" represents a dispersion of colloidal or polymeric particles within a solvent, while the "**gel**" is a tridimensional continuous network containing a liquid phase. The "gel" is obtained due to the chains between the molecular chains while the particles are interacting through weak bonds (e.g. Van der Waals or hydrogen bond).

While the "gel" is dried directly through evaporation, the capillary forces are leading to a contraction resulting a **xerogel**, meaning a solid with a large surface and high porosity. The pores in this case do exhibit dimensions of tens of nanometers order. Case the drying is controlled, through pressure or heating, the solid formed will incorporate bigger pores with low density. These materials are known as **aerogel**.

Thus, in spite of the fact that the method is presuming longer times for synthesis as well as quite high costs for precursors, a series of technological advantages allow the utilization of the method at industrial scale, as:

- simplicity;

- low processability temperature;

- versatility;

- high chemical homogeneity;

- higher purity of the final chemical product;

- possibility of developing 3D polymeric structures;

- minimal investments in manufacturing equipment;

- low consumption energy as the method does not need to reach melting temperatures during the synthesis process;

- thin films development.

A major advantage of the technique is the fact that it can be employed both *in-situ* and *ex-situ*. The in-situ application presumes the fact that the liquid solution can be deposited directly on ceramic or glass support for developing polymeric films.

However, in spite of aforementioned technological advantages, the method demonstrated a significant drawback which is related to the fact that the *homogenization of the dispersed phased onto the polymeric matrix is hampered while the dimensions of the nanoparticles are decreasing due to their aggregation*. A solution for this technological bottleneck is the stabilization of colloidal solutions with the help of electrochemical techniques.

As mentioned earlier, in developing substrates provided with printed layers on top, the contact angle represents an essential technological parameter. The **contact angle** is used for analysing the interaction between the wetting liquid and the solid. Thus, starting from this general definition, the **hydrophobic and hydrophilic surfaces** can be defined.

Hydrophobic surface   **Hydrophilic surface**

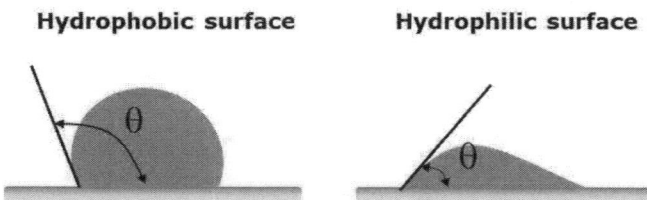

Figure 3.2 – Comparison between contact angle of a the hydrophobic and hydrophilic surfaces

As it can be observed from the above figure, the complete wetting corresponds to a contact angle with a value close to zero, while a lower level of wetting corresponds to a contact angle above 90°.

Starting from this theoretical perspective, it can be said that the value of the contact angle is dependent of the nanocomposite's polarizability. The polarization is the local increase (within a point in the bulk of the material) of the electrical displacement D, fact which is leading to formation of electric dipoles in the presence of an electric field, E, or even in its absence. The dipoles may be formed as of limited and elastic displacement of electric charges from atoms (molecules), of ions in the nodes of crystalline networks or of real dipoles which exist and are randomly distributed in polar objects and which are rotating in the field's direction.

The **dielectric permittivity** ε is used for characterizing the electrical polarization state of a material. Case the permittivity is expressed in relation to vacuum's permittivity ($\varepsilon^0 = 8,85 \times 10^{-12}$F/m) than one can define the relative permittivity as $\varepsilon_r = \varepsilon/\varepsilon_0$

Thus, in respect to all these presented above, the nanocomposites can be categorized as strong polar nanodielectrics and weak polar nanodielectrics.

Strong polar nanodielectrics are those materials which are "getting wet", meaning that their contact angle is above 90°, while the weak polar nanocomposites are not "getting wet", exhibiting a contact angle below 90°.

Within strong polar nanocomposites the dipolar (orientational) polarization is much evident as they are the result of unsymmetrical covalent bond fact which is allowing formation of real electric dipoles in the absence of electric field. The rotation of electric dipoles in the presence of electric fields is leading to dipolar polarization. The magnitude of dipoles' rotation is associated to dipolar molecular moment m. Thus, can be said that at macroscopic level, the contribution of dipolar polarization on the value of the dielectric constant can be attributed to dipolar molecular moment.

The dipolar polarization is occurring slowly, its relaxation time τ is getting values of $10^{-3}...10^{-6}$s, being strongly affected by temperature and frequency, with high energy consumption (high losses).

As a consequence of all those above presented, the characterization of a solid surface can be realized through calculation of surface's free energy and all its components, dispersive and polar, from Owens and Wendt equations. Orientation of hydrophilic-hydrophobic groups at the interface solid-air can be influenced by polar (γ_{sv}^d) and dispersive (γ_{sv}^p) factors of superficial energy (γ_{sv}).

The polar and dispersive component can be calculated with Owens-Wendt equation as follows:

$$(1+cos\theta)\ \gamma_L/(2\ \gamma_L^d) = (\gamma_s^d)^{1/2} + (\gamma_s^p)^{1/2}\ (\gamma_L^p/\gamma_L^d)^{1/} \qquad (3.1)$$

where θ is the contact angle determined for two liquids; γ_s^d and γ_s^p represent the dispersive component and the polar component respectively. The overall surface free energy can be expressed as:

$$\gamma_s = \gamma_s^p + \gamma_s^d \qquad (3.2)$$

Dupre provided the calculus formula of the adhesion between a liquid and a solid, namely the needed energy for separating two surfaces in contacted, W_A:

$$W_A = \gamma_{sv} + \gamma_{lv} - \gamma_{sl} \qquad (3.3)$$

This energy can be also calculated with Young-Dupree equation:

$$W_A = \gamma_{lu}(1+cos\theta) \qquad (3.4)$$

The S_c parameter can be determined from Dupre equation:

$$S_c = \gamma_{su} - \gamma_{sl} - \gamma_{lu} \qquad (3.5)$$

where γ_{sl} can be written as:

$$\gamma_{sl} = \gamma_{su} + \gamma_{lu} - W_A \qquad (3.6)$$

Girifalco and Good had introduced the interaction parameter Φ between the surface and the liquid, and calculated it as:

$$\Phi = \frac{\gamma_{lv}(1 + cos\theta)}{2(\gamma_{lv}\gamma_{sv})^{1/2}} \tag{3.7}$$

According to Young-Dupre equation, the value of W_A is dependent on superficial tension of the liquid and on the contact angle. Thus, the values for W_A are increasing when the contact angle is decreasing. Generally, a positive value of S_c is meaning the substrate will get wet spontaneously and will spread on a solid surface, while case S_c will get a negative value, the wetting will be partial with a limited spreading.

III.2 Conventional polymers employed as matrix for nanocomposite development

The majority of nanocomposite structures used for developing flexible substrates are employing a classical polymer: a matrix onto which various nanoparticles are incorporated for tailoring the physical, chemical and electric properties of the final product. For being used as flexible polymeric substrate, the material should exhibit a series of properties which will ensure proper development of printed structure at its surface.

Thus, in respect to chemical characteristics, the material employed should be capable to free all residual chemical compounds employed during the synthesis process. The surface of the substrate should work as a barrier against water from surrounding ambient as well as against various gases. Moreover, the surface should ensure a high degree of adherence of the ink (e.g. dielectric and/or conductive). Thus, preferably, on short distances, the surface of the material should present less pores, but in the same time, on longer distances, eventual asperities and a high porosity would be needed.

Regarding the thermal properties, the glass transition temperature of the substrate should be correlated with the maximal temperature reached during the printing process in a manner that during the exploitation of the resulted product, deteriorations of the substrate or ink removal should be avoided. Moreover, the dimensional stability of the substrate should be considered as well during the design phase.

In what it concerns the mechanical properties, special attention should be dedicated to material's stiffness for ensuring higher flexibility through a higher Young's modulus.

Regarding the electrical properties, the resistivity of the material should be as lower as possible, coupled with lower dielectric constant and dissipation/losses factor. Moreover, for avoiding the deterioration, as the thickness of the material is very low, the dielectric breakdown should be as larger has possible. From the viewpoint of electromagnetic shielding properties, would be preferable that with the help of conducting nanoparticles (e.g. nanotubes) as many interfaces as possible to be created between the matrix and the fillers in order to facilitate development of intrinsic pseudo-microcapacitors.

In respect to optical properties, the substrate should be capable of transmitting light in the same manner as the glass, thus, they should be as less opaque as possible. Of course, there are applications where the opacity is not affecting the device's functionality (e.g. sensors) and thus, the optic properties are not a priority.

Within the most widespread polymeric structures employed in nanocomposite development, one can mention:

1. Semi-crystalline thermoplastic polymers: PET (Polyethylene terephthalate), HS-PET (heat-stabilized PET), PEN (Polyethylene naphthalate).

2. Non-crystalline thermoplastic polymers: PC (Polycarbonate), PES (polyethersulfone).

3. Polymers with high glass transition temperature: PI (Polyimide), PAR (Polyarylate) and ITO (Indium tin oxid).

III.3. Semi-crystalline thermoplastic polymers

PET (Polyethylene terephthalate) is a semi-crystalline or amorphous thermoplastic of high or low density composed of linear molecules. The PET is obtained from ethylene glycol, dimethyl terephthalate or terephthalic acid.

The PET is a rough (non-elastic) solid material, with good dimensional stability and low hygroscopicity (about 0.4%) and very good stability at light. Besides, the PET is characterized by a very good resistance to chemical substances (including alkaline) and low permeability to gases.

Thermal characteristics of PET

Specific heat ($J\ K^{-1}\ kg^{-1}$)	1200-1350
Coefficient of thermal expansion ($x10^{-6}\ K^{-1}$)	20-80
Thermal conductivity @ 23°C ($W\ m^{-1}\ K^{-1}$)	0,15-0,4
Maximum working temperature (°C)	115 to 170
Minimum working temperature (°C)	-40 to -60

Physical characteristics of PET

Hygroscopicity (%)	<0,7
Hygroscopicity during 24 hours (%)	0,1
Density (g/cm^3)	1,3-1,4
Refractive index	1,58-1,64

Electrical characteristics of PET

Volume resistivity (Ωcm)	10^{16}
Dielectric constant	3,4
Dissipation factor (losses)	0,002
Dielectric breakdown (V/mm)	400

From the aspect viewpoint, the PET is available as transparent and colourless, but often it is opaque and whitish. From the point of view of electrical properties, the PET is a polar material, due to double bond C=O met in the monomer, but its properties can vary drastically from one producer to another.

PET has been used in development of printed electronics. Park used roll-to-toll (R2R) engraving to develop on PET plastic foil, using four types of nanoparticle inks, the components of a wireless sensor meaning the antenna, diode and capacitor with a print speed of 8m / min. The printed antenna exhibited a wireless transmission rate of a 0.3W signal received from a standard 13.56 MHz transmitter.

Moreover, Park manufactured a Cu substrate on PET synthesized with a super-hydrophilic surface containing nanostructures of large thicknesses and heights. The manufactured substrate exhibited a peel strength of $1300Nm^{-1}$ in absence of an intermediate layer, without denoting an increase in electrical resistance after repeated bending cycles with a 1 mm curve height to over 300000 repeated bendings. Park demonstrated the possibility of producing flexible metal / polymer substrates for printed electronics exhibiting excellent interfacial bonding voltages between metal and polymer and a high fatigue resistance produced by repeated bendings.

Potyrailo employed roll-to-roll (R2R) printing technology to produce battery-less RFID gas sensors with high selectivity developed on flexible PET substrate. The selectivity of the RFID sensors developed was determined by measuring the impedance resonance spectrum followed by a multianalysis of the spectrum characteristics and through correlating these spectral characteristics with the target gas vapor concentrations. Simultaneously, analysis of multiple spectrum factors offered the possibility to avoid ambient environmental interferences. The authors successfully tested the sensors' response to toluene, acetone and ethanol.

Screen-printed wireless sensors developed on PET flexible media were studied by Shi as well. The wireless sensors did not require a battery as were provided with passive wireless data transmission and power capability. Sensors manufactured on a flexible PET substrate were designed based on RFID technology.

Metallic patterns connected via microholes filled with conductive paste were printed on both sides of the polymeric film. The printed tracks for electronic components such as resistors and transistors were developed on one side while the coil for data transmission and the wireless feed was printed on the other one.

PEN (Polyethylene naphthalate) is a semi-crystalline polymer, colourless, which is quite similar with PET from chemical structure viewpoint, but more stable at higher temperature. Moreover, in the case of PEN one can notice a lower contraction at temperature variation, a higher glass transition temperature, simultaneously with a higher permeability to oxygen and carbon dioxide. Besides, in comparison to PET, the electrical properties of PEN are superior, as its dielectric breakdown is superior to that of PET with at least 25%. However, the manufacturing cost of PEN are much higher.

Thermal characteristics of PEN:

Specific heat ($J K^{-1} kg^{-1}$)	NA
Coefficient of thermal expansion ($\times 10^{-6} K^{-1}$)	20-21
Thermal conductivity @ 23ºC ($W m^{-1} K^{-1}$)	0,15
Maximum working temperature (ºC)	155
Minimum working temperature (ºC)	NA

Physical characteristics of PEN:

Hygroscopicity (%)	0.4
Hygroscopicity during 24 hours (%)	NA
Density (g/cm^3)	1,36
Refractive index	1,58-1,64

Electrical characteristics of PEN:

Volume resistivity (Ωcm)	10^{15}
Dielectric constant	3,2
Dissipation factor (losses)	0,0048-0,005
Dielectric breakdown (V/mm)	400

Aliane presented the development of improved temperature sensors printed on large surfaces of flexible substrates. Production flow is achieved through printing technology that uses exclusively colloidal solutions. The printed temperature sensors were obtained based on a resistive paste integrated into a Wheatstone bridge. The thickness of the PEN layer was of 125µm. The sensors exhibited satisfactory electrical properties with a sensitivity of 0.06V/°C at a V_{in} = 4.8 V voltage. Sensitivity was enhanced by heat treatment and O_2 plasma treatment.

Ana Moya employed inkjet printing technology as an alternative to standard micro-manufacturing techniques in the field of microsensors on flexible substrates. The PEN substrate proved to be a promising candidate for disposable sensors as of the low production cost. The authors have also proposed the development of all-in-printed sensor by inkjet printing by using commercially available inks that can be modified and sintered at low temperatures of 120 ° C. These inks allow PEN to be used as a flexible substrate and are used to develop amperometric sensors for oxygen detection based on metallic nanoparticles as a sensitive element.

Oh developed a diagnostic system for the rapid detection of bacterial pathogens based on a magnetoresistive (MR) sensor interacting with magnetic micro beads coated with specific antibodies. The magnetoresistive sensor coated by a teflon layer for passivation was manufactured on an organic substrate, which is flexible and cost-effective, while the role of teflon passivation was to maintain its flexibility. A PEN substrate of 280 μm thickness has been used as a support film as it exhibited excellent resistance to solvents as well as very good resistance to mechanical stress and low thermal contraction.

HS-PET (heat-stabilized PET) is another polymeric material derived from PET developed especially as an alternative for classical PET. It is available on the market under the commercial name of Melynex. HS-PET is very stable from dimensional viewpoint up to 150°C as beside the thermal treatment it is undergoing also a UV treatment. The UV treatment is employed for improving the adhesion of various conducting deposits at material's surface. It is available on the market as films with a thickness from 50um to 250um. In what it concerns the production costs, HS-PET is much more costly of being processed than PEN and PET.

Thermal characteristics of HS-PET:

Specific heat ($J\ K^{-1}\ kg^{-1}$)	NA
Coefficient of thermal expansion ($x10^{-6}\ K^{-1}$)	6
Thermal conductivity @ 23°C ($W\ m^{-1}\ K^{-1}$)	NA
Melting point (°C)	265

Minimum working temperature (°C)	150

Physical characteristics of HS-PET:

Hygroscopicity (%)	Inferior to PET
Hygroscopicity during 24 hours (%)	NA
Density (g/cm^3)	Close to PET
Refractive index	Superior to PET

Electrical characteristics of HS-PET:

Volume resistivity (Ωcm)	10^{16}
Dielectric constant	2.9
Dissipation factor (losses)	0.002
Dielectric breakdown (V/mm)	3000

As stated also within the previous paragraphs, Khan and Lorenzelli performed an in-depth study on various printing materials used for both the substrate and the electronic materials (materials for patterning). Different printing methods have been studied (solution or dry printing, contact or noncontact printing) in terms of technology materials and current state of the art development.

The key challenges in different printing techniques and potential application directions were also highlighted. The ability to combine different printing methods had been exploited to ensure technology transfer from laboratory research to roll-to-roll production. The semi-crystalline structure of the HS-PET substrate has attracted attention based on its glass transition temperature higher than 140°C fact which allows polymer casting without significant degradation.

Rand and Richter investigated various materials that can be used as substrates for the production of flexible printed photovoltaic cells. Tests for deformation of the substrate under a heat source (over 30 minutes) exhibited that HS-PET compared to PET is more stable at high temperature as of the pretreatment for thermal stabilization which increases the dimensional stability of the film.

Zardetto tested materials used for developing substrates for flexible printed circuits including HS-PET (thermally stabilized PET). The tests performed were designed to determine the optical parameters, mechanical flexibility (bending under different tension and compressive stresses), resistance to solvent action thermal stability (conductivity and deformation) and UV light irradiation. The results demonstrated the limitations and advantages of using HS-PET as a substrate for printing electronic devices and offered a series of solutions for improving performance. On the other hand, the same study is précising that HS-PEN is more stable than HS-PET because of the naphthalene rings which are increasing the glass transition of the material.

III.4. Non-crystalline thermoplastic polymers

PC (Polycarbonate) is a colourless thermoplastic containing carbonate groups remarkable as of its high resistance to physical impact at negative temperatures and very good resistance to UV. In spite of its very good thermal resistance, dimensional stability and of the fact that can be bent at ambient temperature, PC is not so resistant to chemical agents. From the viewpoint of electrical properties, this polymer is inferior to PET and PEN as it can be noticed from the data presented within the below tables.

Thermal characteristics of PC:

Specific heat ($J\ K^{-1}\ kg^{-1}$)	Approx. 1200
Coefficient of thermal expansion ($\times 10^{-6}\ K^{-1}$)	66-70
Thermal conductivity @ 23°C ($W\ m^{-1}\ K^{-1}$)	0.19-0.22
Maximum working temperature (°C)	115 (max. 130)
Minimum working temperature (°C)	-40

Physical characteristics of PC:

Hygroscopicity (%)	<0.7
Hygroscopicity during 24 hours (%)	0.1
Density (g/cm^3)	1.2
Refractive index	1.5-1.6

Electrical characteristics of PC:

Volume resistivity (Ωcm)	10^{14}-10^{16}
Dielectric constant	2.9
Dissipation factor (losses)	0.01
Dielectric breakdown (kV/mm)	15-67

Chen conducted a study on nanoelectroforming techniques, hot embossing and electrochemical deposition to develop disposable nanostructured biosensors with low cost and high sensitivity. A modified anodic aluminum oxide layer was used to make a thin nickel film deposition pattern. After the corrosion of the aluminum oxide anodic template a three-dimensional mold of the nanostructure was formed. The three-dimensional nickel matrix was further used as a casting replica on a polycarbonate substrate by hot embossing. A gold film was then sputtered over the polycarbonate layer to form the electrode followed by deposition of a uniform layer of gold nanoparticles over the three-dimensional gold electrode by electrochemical deposition. Finally, silver nanoparticles were deposited to improve the conductivity of the sensor. The concentration of the analyte was determined by spectroscopic electrical impedance measurements.

Nanoparticles of MWCNTs printed on polycarbonate substrate by flexographic printing and potential carbon film applications as a sensitive electrochemical element were examined by Tsierkezos. To determine the influence of nanotube surface defects on the electrochemical response, two types of carbon nanotubes based ink were developed. The obtained results demonstrate that defects in the structure of carbon nanotubes improve the electrochemical response detection capability and sensitivity of carbon nanotubes printed on flexible substrate.

Tahara developed a multichannel sensing chip used in a portable device for detecting different tastes. The chip consists of a set of Ti / Ag electrodes coated with a lipid / polymer membrane and a reference electrode. The electrodes were patterned on a polycarbonate substrate. The polycarbonate sheet was provided with a number of holes to ensure access to the electrode and thus to form multiple channels for detection. Dalvand prepared for the first time ZnO nano-needles uniformly distributed on polycarbonate substrate by an electric field assisted chemical method. Superficial morphologies of ZnO nanoscale at different electric fields were studied using scanning and transmission electron microscopy. X-ray diffraction of ZnO nanoparticles showed a highly oriented crystal structure on the c axis. The ZnO nano-scale photoluminescence measurements have identified a peak in the ultraviolet spectrum at 380 nm which is comparable to those which are found in high-quality ZnO films. The increase in ZnO nano-needles on the PC substrate showed the best absorption improvement and best antireflective properties indicating that the product has a great potential for energy recovery devices and photovoltaic cell applications.

PES (polyethersulfone) is an amorphous polymer, transparent, with very good resistance to heat, dimensional stability as well as flame resistance.

Moreover, PES is lightweight and demonstrated good resistance to physical impact, good electrical properties (in particular volume resistivity). However, PES has a relative high hygroscopicity and limited resistance to various chemical agents. Besides, is presenting relative high manufacturing costs, and the processing temperatures are in general very high.

Thermal characteristics of PES:

Specific heat ($J\ K^{-1}\ kg^{-1}$)	1.1
Coefficient of thermal expansion ($x10^{-6}\ K^{-1}$)	50-55

Thermal conductivity @ 23°C (W m^{-1} K^{-1})	0.13-0.18
Maximum working temperature (°C)	Up to 220
Minimum working temperature (°C)	-110

Physical characteristics of PES:

Hygroscopicity (%)	2.2
Hygroscopicity during 24 hours (%)	0.4-1.85
Density (g/cm^3)	Approx. 1.35
Refractive index	1.6-1.7

Electrical characteristics of PES:

Volume resistivity (Ωcm)	10^{17}-
Dielectric constant	3.7
Dissipation factor (losses)	0.003
Dielectric breakdown (kV/mm)	16

3. Polymers with high glass transition temperature

PI (Polyimide) is often met under the commercial name of Kapton and is a polymer with the best electrical properties as of its low dielectric constant at 50Hz (about 2.2) as well as very low dielectric losses (approx. 0.0015-0.002). Moreover, these outstanding electric properties should be corroborated with the fact that the glass transition temperature of PI is above 250°C. PI is a polymer containing mainly aromatic groups with a very good thermal dimensional stability, being insoluble and infusible, with high resistance to UV, and in the majority of cases, is transparent. Its transparency may be lost case the thickness is getting thicker. PI has an outstanding resistance to flame as well as to various chemical agents, but its hygroscopicity is quite high.

Thermal characteristics of PI:

Specific heat (J K^{-1} kg^{-1})	1090

Coefficient of thermal expansion ($\times 10^{-6}$ K^{-1})	30-60
Thermal conductivity @ 23ºC (W m^{-1} K^{-1})	0.1-0.35
Maximum working temperature (ºC)	250-320
Minimum working temperature (ºC)	-250

Physical characteristics of PI:

Hygroscopicity (%)	<0.7
Hygroscopicity during 24 hours (%)	0.2-0.6
Density (g/cm^3)	1.4
Refractive index	1.66

Electrical characteristics of PI:

Volume resistivity (Ωcm)	10^{18}
Dielectric constant	2.2
Dissipation factor (losses)	0.0018
Dielectric breakdown (kV/mm)	220

PAR (Polyarylate) are polymers less utilized as they are produced in small quantities as of their high production costs. This category of polymers is presenting a high crystallinity, very good mechanical properties, very good chemical and heat resistance as well as very high thermal stability.

As an example, Vectran, one of the most known aromatic polyesters, is 5 times more resistant that stainless steel.

Thermal characteristics of PAR:

Specific heat (J K^{-1} kg^{-1})	1.1
Coefficient of thermal expansion ($\times 10^{-6}$ K^{-1})	50-55
Thermal conductivity @ 23ºC (W m^{-1} K^{-1})	0.13-0.18
Maximum working temperature (ºC)	Up to 220

Minimum working temperature (°C)	-110

Physical characteristics of PAR:

Hygroscopicity (%)	2.2
Hygroscopicity during 24 hours (%)	0.4-1.85
Density (g/cm³)	Approx. 1.35
Refractive index	1.6-1.7

Electrical characteristics of PAR:

Volume resistivity (Ωcm)	10^{17}-
Dielectric constant	3.7
Dissipation factor (losses)	0.003
Dielectric breakdown (kV/mm)	16

In respect to all these materials, Khan is providing a comparison of the glass transition temperatures of polymeric materials used as flexible substrate. As it can be observed, an increase of the values for PET and PI is noticed.

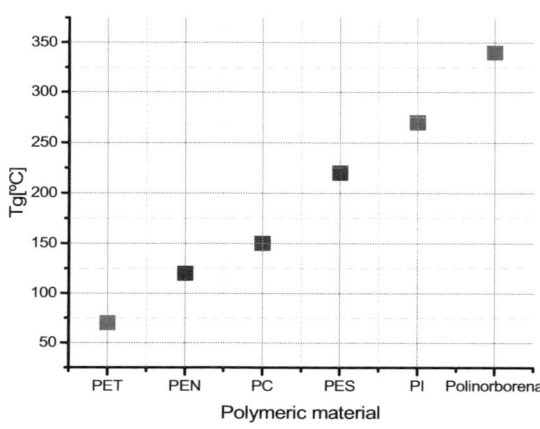

Figure 3.3 – Glass transition temperature for a series of classical polymers employed as flexible substrate

Bibliography

[1] Sang Jin Park, Tae-Jun Ko, Juil Yoon, Myoung-Woon Moon, Kyu Hwan Oh, Jun Hyun Han, Highly adhesive and high fatigue-resistant copper/PET flexible electronic substrates, Applied Surface Science, Volume 427, Part B, 2018, Pages 1-9, ISSN 0169-4332, https://doi.org/10.1016/j.apsusc.2017.08.195.

[2] C. W. P. Shi, Xuechuan Shan, G. Tarapata, R. Jachowicz, J. Weremczuk, H. T. Hui, Fabrication of wireless sensors on flexible film using screen printing and via filling, Microsyst. Technol. 0946-7076, 661-667, 2011, 10.1007/s00542-010-1161- 2

[3] Organic Solar Cells: Fundamentals, Devices, and Upscaling 1st Edition by Barry P. Rand (Editor), Henning Richter (Editor), ISBN-13: 978-9814463652, ISBN-10: 9814463655

[4] Zardetto, V., Brown, T. M., Reale, A. and Di Carlo, A. (2011), Substrates for flexible electronics: A practical investigation on the electrical, film flexibility, optical, temperature, and solvent resistance properties. J. Polym. Sci. B Polym. Phys., 49: 638–648. doi:10.1002/polb.22227

[5] A. Aliane, V. Fischer, M. Galliari, L. Tournon, R. Gwoziecki, C. Serbutoviez, I. Chartier, R. Coppard, Enhanced printed temperature sensors on flexible substrate, Microelectronics Journal, Volume 45, Issue 12, 2014, Pages 1621-1626, ISSN 0026-2692, https://doi.org/10.1016/j.mejo.2014.08.011.

[6] Ana Moya, Enrico Sowade, Francisco J. del Campo, Kalyan Y. Mitra, Eloi Ramon, Rosa Villa, Reinhard R. Baumann, Gemma Gabriel, All-inkjet-printed dissolved oxygen sensors on flexible plastic substrates, Organic Electronics,Volume 39, 2016, Pages 168-176, ISSN 1566-1199, https://doi.org/10.1016/j.orgel.2016.10.002.

[7] Sunjong Oh, Mital Jadhav, Jaein Lim, Venu Reddy, CheolGi Kim, An organic substrate based magnetoresistive sensor for rapid bacteria detection, Biosensors and Bioelectronics, Volume 41, 2013, Pages 758-763, ISSN 0956-5663, https://doi.org/10.1016/j.bios.2012.09.069.

[7] Marcel Štofik, Alena Semerádtová, Jan Malý, Zdeňka Kolská, Oldřich Neděla, Dominika Wrobel, Petr Slepička, Direct immobilization of biotin on the micro-patterned PEN foil treated by excimer laser, Colloids and Surfaces B: Biointerfaces, Volume 128, 2015, Pages 363-369, ISSN 0927-7765, https://doi.org/10.1016/j.colsurfb.2015.02.032.

[8] Chen Y-S, Wu C-C, Tsai J-J, Wang G-J. Electrochemical impedimetric biosensor based on a nanostructured polycarbonate substrate. International Journal of Nanomedicine. 2012;7:133-140. doi:10.2147/IJN.S27225.

[9] Tsierkezos, Nikos G.; Wetzold, Nora; Hubler, Arved Carl; Ritter, Uwe; Szroeder, Paweł ,Sensor Letters, Volume 11, Number 8, August 2013, pp. 1465-1471(7), American Scientific Publishers, DOI: https://doi.org/10.1166/sl.2013.2987

[10] Y. Tahara, Y. Maehara, J. Ke, A. Ikeda and K. Toko, "Development of a multichannel taste sensor chip for a portable taste sensor," 2012 IEEE Sensors, Taipei, 2012, pp. 1-4, doi: 10.1109/ICSENS.2012.6411197

[11] Ramazanali Dalvand, Shahrom Mahmud, Jalal Rouhi, C.H. Raymond Ooi, Well-aligned ZnO nanoneedle arrays grown on polycarbonate substrates via electric field-assisted chemical method, Materials Letters, Volume 146, 2015, Pages 65-68,ISSN 0167-577X, https://doi.org/10.1016/j.matlet.2015.02.003.

[12] Petr Juřík, Petr Slepička, Zdeňka Kolská, Václav Švorčík, Change of surface properties of gold nano-layers deposited on polyethersulfone film due to annealing, Materials Letters, Volume 165, 2016, Pages 33-36, ISSN 0167-577X, https://doi.org/10.1016/j.matlet.2015.11.098.

[13] Liyakat Hamid Mujawar, Iqbal M.I. Ismail, Zulfiqar Ahmad Rehan, Mohammad Soror El-Shahawi, A miniaturized assay for sensitive determination of Cu2+ ions on nanolitre arrayed 4-(2-pyridylazo)resorcinol (PAR) spots on polyethersulfone membrane platform, Journal of Molecular Liquids, Volume 229, 2017, Pages 574-582, ISSN 0167-7322, https://doi.org/10.1016/j.molliq.2016.12.085.

[14] Inkyu Lee, Wan-Kyu Oh, Jyongsik Jang, Screen-printed fluorescent sensors for rapid and sensitive anthrax biomarker detection, Journal of Hazardous Materials, Volumes 252–253, 2013,Pages 186-191, ISSN 0304-3894, https://doi.org/10.1016/j.jhazmat.2013.03.003.

[15] Trung, T. Q., Tien, N. T., Kim, D., Jang, M., Yoon, O. J. and Lee, N.-E. (2014), A Flexible Reduced Graphene Oxide Field-Effect Transistor for Ultrasensitive Strain Sensing. Adv. Funct. Mater., 24: 117–124. doi:10.1002/adfm.201301845

[16] Gun Woo Hyung, Jaehoon Park, Ja Ryong Koo, Kyung Min Choi, Sang Jik Kwon, Eou Sik Cho, Yong Seog Kim, Young Kwan Kim, ZnO thin-film

transistors with a polymeric gate insulator built on a polyethersulfone substrate, Solid-State Electronics, Volume 69, 2012, Pages 27-30, ISSN 0038-1101, https://doi.org/10.1016/j.sse.2011.12.001.

[17] Jin-Jin Wang, Hong Hu, Cheng-Huo Shang, Effect of annealing on the performance of nickel thermistor on polyimide substrate, Thin Solid Films, Volume 632, 2017, Pages 28-34, ISSN 0040-6090, https://doi.org/10.1016/j.tsf.2017.04.034.

[18] E. Skotadis, D. Mousadakos, K. Katsabrokou, S. Stathopoulos, D. Tsoukalas, Platinum Nanoparticle Chemical Sensors on Polyimide Substrates, Procedia Engineering, Volume 47,2012, Pages 778-781, ISSN 1877-7058, https://doi.org/10.1016/j.proeng.2012.09.263.

[19] Ayda Bouhamed, Alireza Mohammadian Kia, Slim Naifar, Volodymyr Dzhagan, Christian Müller, Dietrich R.T. Zahn, Slim Choura, Olfa Kanoun, Tuning the adhesion between polyimide substrate and MWCNTs/epoxy nanocomposite by surface treatment, Applied Surface Science, Volume 422, 2017, Pages 420-429, ISSN 0169-4332, https://doi.org/10.1016/j.apsusc.2017.05.177.

[20] Mathilde Rieu, Malick Camara, Guy Tournier, Jean-Paul Viricelle, Christophe Pijolat, Nico F. de Rooij, Danick Briand, Fully inkjet printed SnO2 gas sensor on plastic substrate, Sensors and Actuators B: Chemical, Volume 236, 2016, Pages 1091-1097, ISSN 0925-4005, https://doi.org/10.1016/j.snb.2016.06.042.

[21] Geonwook Yoo et al., Flexible and Wavelength-Selective MoS2 Phototransistors with Monolithically Integrated Transmission Color Filters, Scientific Reports 7, Article number: 40945 (2017), doi:10.1038/srep40945

[22] Linrun Feng, Chen Jiang, Hanbin Ma, Xiaojun Guo, Arokia Nathan, All ink-jet printed low-voltage organic field-effect transistors on flexible substrate, Organic Electronics, Volume 38, 2016, Pages 186-192, ISSN 1566-1199, https://doi.org/10.1016/j.orgel.2016.08.019.

[23] Yong-Hoon Kim et al., Flexible metal-oxide devices made by room-temperature photochemical activation of sol–gel film, Nature 489, 128–132 (06 September 2012), doi:10.1038/nature11434

[24] A. Renitta, K. Vijayalakshmi, High performance hydrogen sensor based on Mn implanted ZnO nanowires array fabricated on ITO substrate, Materials

Science and Engineering: C, Volume 77, 2017, Pages 245-256, ISSN 0928-4931, https://doi.org/10.1016/j.msec.2017.03.234

[25] Ramendra K. Pal, Subhas C. Kundu, Vamsi K. Yadavalli, Biosensing using photolithographically micropatterned electrodes of PEDOT:PSS on ITO substrates, Sensors and Actuators B: Chemical,

Volume 242, 2017, Pages 140-147, ISSN 0925-4005, https://doi.org/10.1016/j.snb.2016.11.049.

[26] A. Renitta, K. Vijayalakshmi, Highly sensitive hydrogen safety sensor based on Cr incorporated ZnO nano-whiskers array fabricated on ITO substrate, Sensors and Actuators B: Chemical, Volume 237, 2016, Pages 912-923, ISSN 0925-4005, https://doi.org/10.1016/j.snb.2016.07.017.

[27] Lijun Ding, Yan Gao, Junwei Di, A sensitive plasmonic copper(II) sensor based on gold nanoparticles deposited on ITO glass substrate, Biosensors and Bioelectronics, Volume 83, 2016, Pages 9-14, ISSN 0956-5663, https://doi.org/10.1016/j.bios.2016.04.002.

[28] Camila de L. Ribeiro, João Guilherme M. Santos, Jurandir R. de Souza, Marcelo A. Pereira-da-Silva, Leonardo G. Paterno, Electrochemical oxidation of salicylic acid at ITO substrates modified with layer-by-layer films of carbon nanotubes and iron oxide nanoparticles, Journal of Electroanalytical Chemistry, Volume 805, 2017, Pages 53-59, ISSN 1572-6657, https://doi.org/10.1016/j.jelechem.2017.09.063.

[29] Siqi Li, Pengfei Lin, Liupeng Zhao, Chong Wang, Deye Liu, Fangmeng Liu, Peng Sun, Xishuang Liang, Fengmin Liu, Xu Yan, Yuan Gao, Geyu Lu, The room temperature gas sensor based on Polyaniline@flower-like WO3 nanocomposites and flexible PET substrate for NH3 detection, Sensors and Actuators B: Chemical, Volume 259, 2018, Pages 505-513, ISSN 0925-4005, https://doi.org/10.1016/j.snb.2017.11.081.

[30] Hsin-Cheng Lai, Zingway Pei, Jyun-Ruri Jian, and Bo-Jie Tzeng, Alumina nanoparticle/polymer nanocomposite dielectric for flexible amorphous indium-gallium-zinc oxide thin film transistors on plastic substrate with superior stability, Appl. Phys. Lett. 105, 033510 (2014); https://doi.org/10.1063/1.4891426

[31] Raj, P.M., Sharma, H., Reddy, G.P. et al. Journal of Elec Materi (2014) 43: 1097., DOI:https://doi.org/10.1007/s11664-014-3025-5

[32] Arao Y. (2015) Flame Retardancy of Polymer Nanocomposite. In: Visakh P., Arao Y. (eds) Flame Retardants. Engineering Materials. Springer, Cham, DOI https://doi.org/10.1007/978-3-319-03467-6_2

[33] A. S. M. Alqadami and M. F. Jamlos, "Compact and conformal multilayer antenna based on polymer nanocomposite substrate," 2015 IEEE International RF and Microwave Conference (RFM), Kuching, 2015, pp. 180-182. doi: 10.1109/RFM.2015.7587739

[34] Ruimin Ma, Xiao Wan, Teng Zhang, Nuo Yang, Tengfei Luo, Interfacial Thermal Transport in Boron Nitride-Polymer Nanocomposite, arXiv:1711.11201

[35] Tongfan Hao, Zhiping Zhou, Yijing Nie, Ya Wei, Zhouzhou Gu, Songjun Li, Effect of the polymer-substrate interactions on crystal nucleation of polymers grafted on a flat solid substrate as studied by molecular simulations, Polymer, Volume 123, 2017, Pages 169-178, ISSN 0032-3861, https://doi.org/10.1016/j.polymer.2017.07.020.

[35] Barbara Horváth, Jin Kawakita, Toyohiro Chikyow, Adhesion of silver/polypyrrole nanocomposite coating to a fluoropolymer substrate, Applied Surface Science, Volume 384, 2016, Pages 492-496, ISSN 0169-4332, https://doi.org/10.1016/j.apsusc.2016.05.079.

[36] Mengru Li, Wei Wang, Zhuo Chen, Zhiling Song, Xiliang Luo, Electrochemical determination of paracetamol based on Au@graphene core-shell nanoparticles doped conducting polymer PEDOT nanocomposite, Sensors and Actuators B: Chemical, Volume 260, 2018, Pages 778-785, ISSN 0925-4005, https://doi.org/10.1016/j.snb.2018.01.093.

[37] Dongzhi Zhang, Dongyue Wang, Peng Li, Xiaoyan Zhou, Xiaoqi Zong, Guokang Dong, Facile fabrication of high-performance QCM humidity sensor based on layer-by-layer self-assembled polyaniline/graphene oxide nanocomposite film, Sensors and Actuators B: Chemical, Volume 255, Part 2, 2018, Pages 1869-1877, ISSN 0925-4005, https://doi.org/10.1016/j.snb.2017.08.212.

Chapter IV - Dielectrophoretic manipulation of particles and living cells

IV.1. Theory of Dielectroforesis

The **dielectrophoresis (DEP)** is the electrokinetic motion of dielectrically polarized material in a fluid under the action of non-uniform electric fields. The phenomenon was firstly described by Pohl in early nineteen-fifty and can be noticed both in direct current (DC) and in alternative current (AC) as well. In the presence of a non-uniform electric field, the object will be displaced towards or off the stronger field in case its permittivity will be higher or smaller than the permittivity of the suspending fluid. Case the particle is pulled towards the stronger field region, than we can refer to this phenomenon as *positive DEP*. Should the suspended object be repelled, then we can call the effect *negative DEP*. From chronological viewpoint, the principle of DEP is newer than the one of **electrophoresis (EC)**. In the case of **electrophoresis (EC)**, a uniform electric field is facilitating the appearance of equal Columbic forces for each half of the target object and consequently the overall net force imposed on the particle will be zero. EC was firstly developed in 1930 and shortly was improved in the following three decades.

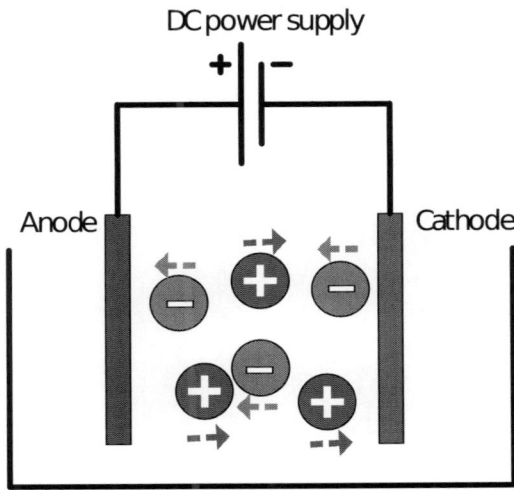

Figure 4.1 – Schematic representation of electrophoresis (EC)

EC assumes the application of a spatially uniform electric field at the level of a net charge which will practically be displaced towards an anode or a cathode depending on its polarity due to *electrophoretic force (F_{EC})*. Should the polarity of the electric change, the displacement direction of particles will change as well. Depending on the magnitude of the F_{EC}, the target object will move till getting in contact with the anode or the cathode respectively. Once the target objects are in contact with one of the electrodes, an exchange of electrons will occur, and thus, the object's trapping may be lost.

For EC, one can define the electrophoretic force which is provided by the formula:

$$\vec{F}_{EC} = Q\vec{E}$$

(4.1)

On the other hand, DEP, as defined above, can be used both in AC and DC but the major difference in comparison to EC is that the electric field gradient is non-uniform. However, DEP does not need the particle to be charged as in the presence of the non-uniform electric field it will act as an electric dipole which will be characterized by an induced dipole moment.

Practically, in the presence of an external electric field the charges will segregate. DEP is strongly dependent on particles' dielectric properties as well as medium polarizability and thus Maxwell-Wagner or interfacial polarization mechanism may manifest.

The interfacial polarization occurs at lower frequencies and, in theory, is corresponding to α-relaxation. Generally, this polarization requires a longer period of time to arise as of the fact that charges have to travel on longer distances in comparison to their dimensions.

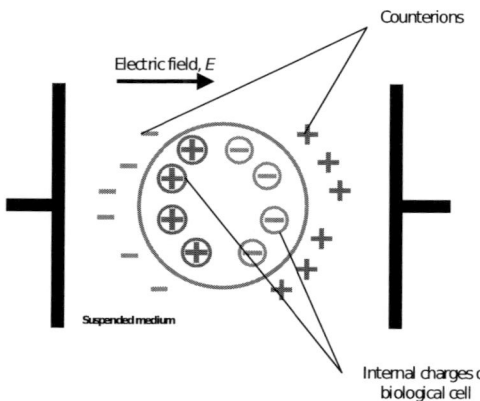

Figure 4.2 – Interfacial or Maxwell-Wagner polarization in case of living cells

In case of living or biological cells, Maxwell-Wagner polarization is dependent on two factors:

1. Heterogeneity of the living cell which is providing the electrical status of the cell which is changing in the presence of the electric field.

2. The counter ions which may form an "electric shield" around the target object or particle due to electric gradient.

The Maxwell-Wagner polarization of a living cell when submitted to electric fields can be schematized as follows:

For understanding the dielectrophoresis, we should firstly consider our object as a spherical particle, suspended within a fluid. Thus, the dielectrophoretic (DEP) force acting at the level of the particle can be expressed as:

$$\vec{F}_{DEP} = 2\pi\varepsilon_m r^3 Re[K(\omega)]\nabla\vec{E}_{rms}^2 \tag{4.2}$$

where r is the particle's radius, ε_m is the permittivity of the medium where the particle is suspended, E_{rms} is the value of the applied field, $Re[K(\omega)]$ is the real part of the Clausius-Mossotti (CM) factor. For a spherical dielectric particle, CM factor can be expressed as:

$$K(\omega) = \frac{\varepsilon_p^* - \varepsilon_m^*}{\varepsilon_p^* + 2\varepsilon_m^*} \tag{4.3}$$

Or, as

$$K(\omega) = \frac{\sigma_p^* - \sigma_m^*}{\sigma_p^* + 2\sigma_m^*} \tag{4.4}$$

where ε_m^* is the complex permittivity of the fluid, ε_p^* the one of the particle, σ_m^* is the complex conductivity of the fluid, σ_p^* is the one of the particle. The object suspended in fluid under the action of the dielectrophoretic force will be attracted or repelled by the regions of higher electric gradient depending on the sign and magnitude of the CM. Should $Re[K(\omega)]>0$, the particle will be displaced towards regions of higher electric gradient, otherwise, the particle is to be displaced towards lower electric gradient regions. As it can be observed from equation 4.2, the dielectrophoretic force is directly proportional with the square electric field fact which is not changing the direction of the particle's movement when the polarity of the electric field is changed.

Furthermore, another polarization which should be consider when performing DEP is the orientational or dipolar polarization. This type of polarization is not related to the target object but to suspending medium which in majority of cases is a polar fluid. The polar fluids contain polar molecules (e.g. water molecule) which are not symmetric and which are presenting a permanent dipole moment. In the absence of the external electric field the permanent dipole moments of the molecules are randomly distributed within the fluid. However, in the presence of the electric field, the molecules will start a rotational movement in the direction of the field leading to an augmentation of the local electric field within the fluid. The dipolar polarization is manifesting very fast with very high losses but at higher frequencies (e.g. in the case of water above GHz). Consequently, DEP at higher frequencies may be obstructed by dipoles rotations.

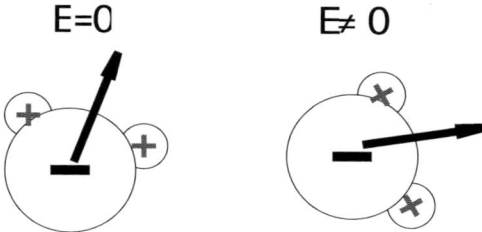

Figure 4.3 – Dipolar or orientation polarization of water molecule

Moreover, the capacity of a DEP system to sort and trap various particles is directly influenced by object's dimensions. As it can be noticed from equation 1, the dielectrophoretic force is dependent on object's volume. Consequently, the larger the particle will be, stronger the F_{DEP} will be as well as under the circumstances that all the other factors of influence will remain constant. One can say that F_{DEP} is a non-linear force (ponderomotive). Implicitly, the object's dielectrophoretic velocity under electric field will be influenced, and thus the levitation height of the particle in respect to the electrodes of the system.

The critical frequency when the positive dielectrophoresis is transformed into negative dielectrophoresis and contrariwise was called cross-over frequency, and can be determined as:

$$f = \frac{1}{2\pi\tau} = \frac{1}{2\pi}\frac{\sigma_p + 2\sigma_m}{\varepsilon_p + 2\varepsilon_m} \tag{4.5}$$

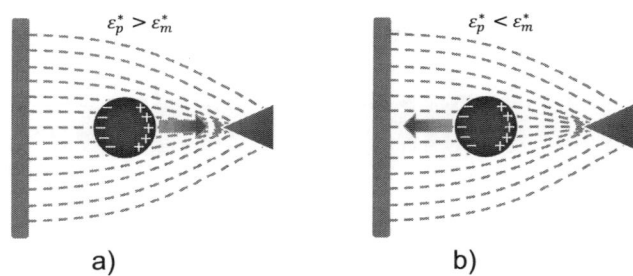

a)   b)

Figure 4.4 – Graphical representation of DEP a) positive and b) negative

where τ is the relaxation time, σ_p is the conductivity of the particle and σ_m is the conductivity of the suspended medium. The cross-over frequency is the frequency point when $Re[K(\omega)]$ is zero.

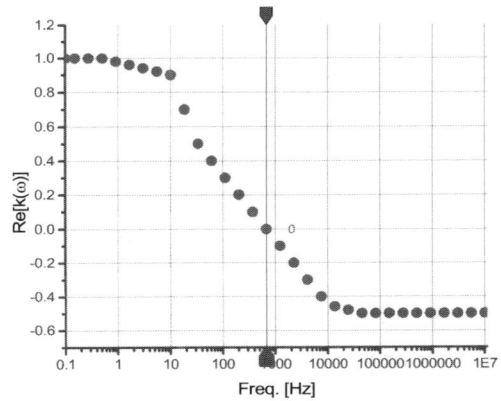

Figure 4.5 – Graphical representation of $Re[K(\omega)]$ vs frequency for an arbitrary target particle

From practical viewpoint, the set-up principle of DEP workbench is quite simple, consisting of a voltage source, a lock-in amplifier and a peristaltic pump. Prior to manipulation, particles must be solubilized while cells must be detached from the extracellular matrix context and dispersed in a physiologic fluid. After particle solubilisation or cell dispersion, the fluid containing particles under test is pumped within a micro-chamber placed on the top of a microfluidic platform. During pumping, AC or DC voltage is applied at the terminals of the microelectrodes. Asymmetry of the working electrodes will lead to occurrence of different sign and value forces producing a displacement of the particles towards one of the electrodes, depending of fluid's and particle's dielectric permittivity, allowing particles to trap between microelectrodes.

The basic DEP set-up is presented in the following picture. Besides the AC voltage, within the depicted DEP set-up a lock-in amplifier is used for monitoring the impedance of trapped objects. The technique is called Dielectrophoretic Impedance Measurement (DEPIM) and is shortly described within the following paragraphs.

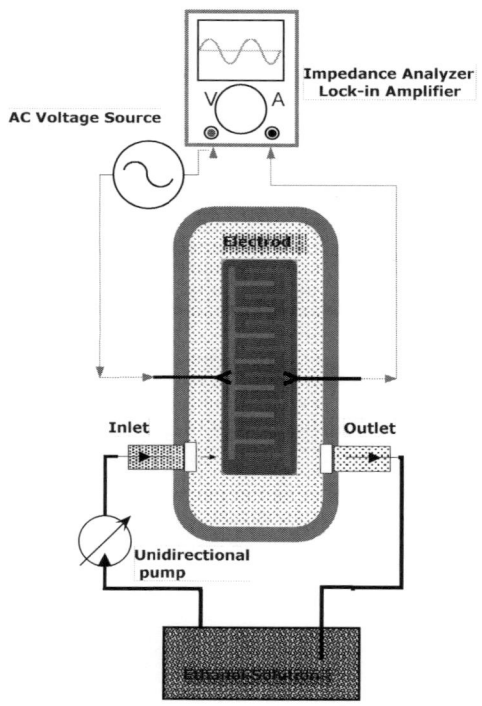

Figure 4.6 – Schematization of DEP set-up

For ensuring operability of a DEP system, the (micro) electrodes' material, shape and dimension play a crucial role. Classical metals have been previously employed for electrodes development as gold (Au), silver (Ag), platinum (Pt), copper (Cu), chromium (Cr), palladium (Pd) and titanium (Ti). Besides these, in the last years, due to technology development carbon (C) and indium tin oxide (ITO) have been reported as suitable materials for interdigitated microelectrodes manufacturing. Moreover, in the last years, due to development of screen-printing methods, conductive silver (Ag) or gold (Au) inks have been demonstrated to be a successful alternative to conventional metallic materials.

In respect to materials used as substrate for electrodes patterning, classical materials as glass, quartz or sapphire have been employed. However, lately, for decreasing the manufacturing cost and for ensuring device's flexibility, a set of polymeric materials have been reported as polydimethylsiloxane (PDMS) or PDMS composite, polymethylmethacrylate (PMMA), poly(ethylene terephthalate) (PET) and polyimide (PI). To explore the benefits and the full potential of objects electromanipulation through DEP, scientist have designed different electrodes geometries shapes in order to optimize the electric gradient generated between the electrodes as well as the DEP force. Optimal electrode design and distribution of current density has been analysed through simulation with specialised FEM based software such as Comsol Multiphisics or Ansys.

The most often met design of microelectrodes' employed in dielectrophoretic manipulation is the interdigitated (micro) electrodes array. The microelectrodes may be castellated or in saw-teeth design. These design have been used extensively for separating, sorting or trapping various types of cells. The geometry, the shape and the topology is significantly influencing electric field gradient, as well as the capacity of the DEP system to control particles or living cells. Moreover, by increasing the number of microelectrodes at the level of a DEP array the sorting and separation capacity of the entire device will be improved.

Another interesting design of electrodes met in literature is the dot electrode geometry which possesses several advantages due to its well-defined and confined region of analysis, effective electric field penetration as well as axisymmetrical electric field distribution. The benefit of this radial electrode geometry, is that it does not require a field mapping or image segmentation to measure the DEP force acting on the target objects. The quantification of the DEP force depends on the change shifts in light transmission through the dot before and after electric potential application.

Quadrupolar electrodes represent a well-established method of manipulation and sorting in both cell and proteins, partly because of their field distribution which can be solved analytically. It is considered that the quadrupole electrode arrangement was derived from setups for electrorotation experiments.

Another direct method to identify the electrical properties of particles by exploiting their dielectrophoretic mobility is to generate an electric field while employing a classical solution with known electrical properties. Particle properties can be determined by measuring their mobility, as particle location can be detected as a function of time from microscopic images.

According to Haapalainen, by using quadrupole electrode geometry, a non-uniform electric field with a linear gradient can be generated. Due to the linearly changing electric field, the DEP force affecting a particle remains constant as the particle moves within the active region of the platform (between the electrode tips). The predetermined mobility behaviour of particles is influenced by the net effect of a constant DEP force and the opposing frictional force of viscous fluid. This behaviour can be used to calculate the permittivity and conductivity of a particle from its mobility in the set magnitude of the electric field gradient, the known viscosity, and the permittivity and conductivity of the carrier fluid.

Should not be neglected the fact that at the level of electrodes, when performing DEP, a third polarization which may occur. It is about **electrode polarization** which may defined as the accumulation of charges at the interface between the suspending medium and the electrodes. As in the case of interfacial polarization, electrode polarization is arising at lower frequencies, in the majority of cases below 10kHz, and is characteristic to strong polar suspensions. Electrode polarization introduce high parasitic capacitances in the measuring set-up and its influence should be removed in the case of performing either dielectric or impedance spectroscopy. The phenomenon was addressed and discussed within many studies but its influence is still challenging.

IV.2. Examples of dielectrophoresis set-ups for nanoparticles and living cells manipulation

To date, considering the high potential of application of DEP in nanoparticles as well as living cells separation and sorting, many technological configurations of DEP were reported by the scientific society:

DEPIM (Dielectrophoretic Impedance Measurement) was developed by Suehiro and is based on positive DEP force to trap living cells onto an interdigitated microelectrode array in the form of pearl-chains. Simultaneously, the electrical impedance is continuously measured while the number of trapped cells is increasing. The method was successfully employed for detection bacteria, norovirus and rotavirus.

DEP-FFF or dFFF (Dielectrophoretic Field-Flow Fractionation), is a DEP method firstly reported in 1997, based on the fact that particles suspended within a fluid may levitate between two electrodes at various distances depending on the polarizability. The difference between particles polarizabilities is provided by the target object's dielectric constants. DEP-FFF was previously applied for separation of cancer cells (e.g. breast cells), electroporated and non electroporated cells, human leukocyte subpopulations.

eDEP (Electrodeless Dielectrophoresis) represents a DEP derived technique which is avoiding the use of metallic (micro)electrodes and their direct contact with the sample. Practically, between the electrodes and the object under DEP force will exist air, but the strong value of the electric field will allow particles manipulation as well. The technique is currently employed for 3D alignment of both living cells and various particles at micro level. Chiou reported in 2015 the concentration of nanoscale particles and proteins in a 150 nm nanoconstriction gap in a microchannel, but eDEP was also used for trapping proteins and DNA.

iDEP (insulator-based DEP) the field gradient is practically tailored with a set of insulator barriers which have an anti-fouling effect. Thus, in the case of iDEP the (micro) electrodes are placed within insulating structure. Higher electric fields may be applied with this technique from the moment that the gas generation drawback met in the case of classical DEP is avoided. The technique is suitable for various biological applications as the method is chemically inert.

twDEP (Travelling Wave Dielectrophoresis), also known as four-phase traveling wave DEP) was developed in 1992 in order to increase the accuracy of trapping and sorting but also motion control and selectivity. The twDEP set up is consisting of multiple discrete electrodes, longer, which are sequentially fed with signals of shifted electric phases, ($0°$, $90°$, $180°$, $270°$). The magnitude of the DEP force is leading to cells levitation as well as parallel displacement along the microelectrodes.

cDEP (contactless DEP) generates non-uniform electric gradients without direct contact between the electrodes and the target object. The electric field is created within microchannels within which the electrodes are sink in conductive solutions and separated from the dielectrophoretic chamber with an insulator. A capacitance will appear between the two electrodes along the microchamber. However, cDEP is limited to application of AC electric gradients which can be also tailored through insulating walls geometry. To date, cDEP was applied at the level of various cancer lines as MDA-MB-231, MCF-7 and THP-1.

oDEP (optically induced dielectrophoresis) demonstrated advantages as low light energy (~ 102 W/cm2), virtual electrodes, increased manipulating area and rapid analysis of cells. Optically-induced-dielectrophoresis (oDEP)-based techniques can be considered a real asset especially for cell manipulation since it allows non-contact cell manipulation and simplicity in microfabrication and operation.

oDEP uses light images as virtual electrodes to generate a non-uniform electric field gradient within the light-illuminated area for electrically manipulating the charged target biological cells. The optically induced dielectrophoresis (ODEP) and fluorescent microscopic imaging in a microfluidic system was used to purify of circulating tumour cells (CTCs) after the conventional CTC isolation methods.

IV.3. Practical utilization of dielectrophoretic nanoparticles and living cells manipulation

DEP has been extensively used before for electromanipulation of nanoparticles and living cells. Suehiro manipulated dielectrophoretically single-wall carbon nanohorns (SWCNHs) and carbon nanotubes (CNTs). Suehiro aimed to exploit the high surface area and high gas-absorption capacity of SWCNHs as sensing element for nitrogen dioxide (NO_2) or ammonia (NH_3) gas in order to develop a gas sensor. The SWCNHs were dispersed in ethanol, (0.2–2.0 mg/ml final concentration), ultrasonicated for 60 min. An ac voltage of 100 kHz frequency and 8 V amplitude (peak to peak value) was employed for trapping at the level of castellated interdigitated microelectrodes while the suspension was continuously pumped through the chamber at 0.5 ml/min. Simultaneously, electrode impedance was continuously measured using a lock-in amplifier in order to monitor the SWCNHs trapping activity. This study followed a study from 2003 of the same research team which manipulated carbon nanotubes (CNTs) using the same configuration of interdigitated microelectrodes. The frequency and the magnitude of the electric ac field were the same as the ones employed in the case of SWCNHs.

Besides, multi-walled carbon nanotubes (MWCNTs) were electrically manipulated with DEP at the level of two triunghiular interdigitated aluminum electrodes with nine fingers. The electric field employed in this case was of a sinusoidal voltage of 10 Vpp at a frequency of 2 MHz. The increase of electric voltage magnitude as well as frequency cannot be attributed to MWCNTs structure but rather to triunghiular geometry of the microelectrodes.

3D numerical modelling and simulation of nanowires manipulated dielectrophoretically was reported by Liu in 2006. Simulation results for the three-dimensional dynamic electric field-guided assembly of NWs in a fluid using the immersed electrokinetic finite was presented. It was highlighted that nanowires are displaced by the DEP force toward the regions of higher electric field. Moreover, case the length of nanowires is close to that of the gap between the microelectrodes, the electric field around the nanowires will be distorted. The maximum of DEP force is obtained when the ratio between the NW length and the gap dimension equals 0.85. This theoretical and numerical statement is supported also by the study of Rotaru and its team which electromanipulated spin crossover nanorods via DEP.

Our team reported for the first time DEP manipulation of onion-like carbon or nano-onions at the level of a classical structure of interdigitated microelectrodes[1]. The AC signal generated by a function generator had a frequency range from 1 kHz to 1 MHz and magnitude ranging from 3 Vpp up to 20 Vpp. The conductance registered a consistent increase until the voltage reached 20 Vpp, then a saturation plateau was noticed. The effects of the AC frequency on DEP were studied by holding the voltage at 10 Vpp and varying the frequency from 0 MHz to 10 MHz.

DEP has been also used to manipulated a series of bioparticles as eukaryotic/prokaryotic cells or cell organelles, small particles as viruses, DNA and proteins. The Meighan's review highlighted that DEP has also been used for separation of various types of pathogen as bacteria, spores and/or viruses from each other or other particle types, "determining viability or life stage of pathogens, or pathogen separation from environmental samples". Different mammalian cells were also characterized using DEP. Mouse stem cells and differentiated offspring were analyzed, and it was determined that the stem cells, differentiated neurons, and differentiated astrocytes possessed varying dielectric properties. DEP was used to concentrate cardiomyocytes to form a cell monolayer over the electrodes, which was followed by impedance measurements. The cardiomyocytes were then treated with endothelin-1; the treated cells had elevated impedance, which was inferred as a strengthening of the cells' attachment to the substrate surface. The study is mentioning that several novel applications of DEP techniques have also been in the last years employed for protein sorting. As an example, Bessette successfully mapped antibody epitopes using an AC DEP microfluidic device, and on the other hand, DEP technique functionalized small particles (880 nm avidin-modified latex) in continuous flow with a "particle exchanger". Besides, DEP papers centered on the manipulation, trapping, sorting, or electrostretching of various-sized DNA, exploring the impact of electroosmotic flow, humidity, and surface conductivity on dielectophoretic sorting, while electrodeless DEP was employed to sort and manipulate DNA based on length-dependent DNA polarizabilities.

In the past years, DEP systems became important in stem cell, cancer or degenerative disease research, while they may act as platforms required to

---

[1] Olariu M, Arcire A and Plonska-Brzezinska M. E., Controlled Trapping of Onion-Like Carbon (OLC) via Dielectrophoresis, Journal of Electronic Materials, Vol. 46, No. 1, 2017, DOI: 10.1007/s11664-016-4870-1.

sort cells or to investigate cell behaviour, according to membrane physical properties. DEP can be used to manipulate biological particles as mammalian cells, yeast, bacteria, viruses, DNA, protein and polystyrene beads. Actually, DEP systems are able to perform even single cell analysis and to investigate intracellular processes and pathways. Biochemical or biocompatibility assays, assessment of drug effects are best developed on single cell based sensors. Various assays assisted by DEP systems, mainly cell sorting, can be completed and supplemented by classic methods as fluorescent activated cell sorting (FACS)or magnetic beads sorting techniques.

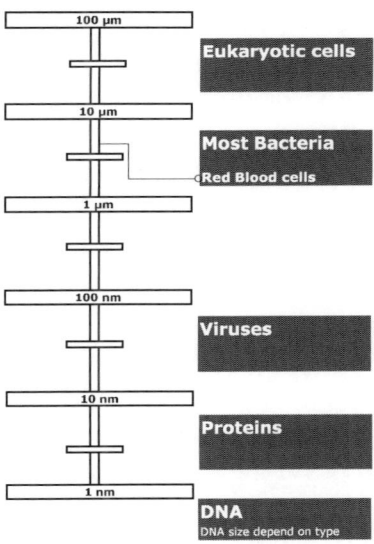

Figure 4.7 – Common biological cells dimensions

In any research or diagnosis platform, one of the most important steps consists in sorting normal from modified cells, the latter category including various changes in morphology of premalignant/malignant cells. Cells transition from normal to pathological state is accompanied by slight detectable changes in its electrical properties indicating the potential use of electrophysiological markers in order to identify these cells. It was demonstrated that electrical properties of normal and malignant cells (tumor cell lines) are varying a lot (malignant breast epithelial cells are different from quasi-normal breast cells, and malignant oral epithelial cells are different from quasi-normal dermalepithelial cells). Yang distinguished an oral squamous cell carcinoma cell line could from an immortalized, non-cancer-derived esophagus cell line by measurements based on a DEP system. It was also demonstrated that important cell functions regarding cancer cells, as adherence and proliferation, generate changes in the cell membrane capacitance and conductivity.

One of the DEP system advantages is that measurements can be performed as fast as in 30 seconds, with cells in suspension, with no requirements for cell adhesion. Simultaneously, cell adhesion and matrix deposition at electrode level in a DEP system can be used to explore adherent cell properties prior and after various treatments or biomaterial biocompatibility testing. The cell conductivity or capacitance can be measured in a DEP-system. Moreover, capacitance per unit area of the membrane depends directly on the membrane surface area, permittivity and thickness. While an entirely smooth cell shows a capacitance of 0.6μF/cm2, cell membrane temporary expansions, as blebs, folds, ruffles or microvilli are increasing these values, creating differences that can be exploited for cell sorting. Moreover, malignant cells show higher capacitance values than dysplastic and normal cells, while increased membrane ruffling characterizes malignant cells and is consistent with the metastatic potential. DEP as a non-invasive technique for detection of oral cancer and oral pre-cancer. As an example, DEP allowed the non-invasive analyses of cytoplasmic conductivity and membrane capacitance of primary normal oral keratinocytes and pre-cancerous and cancerous oral keratinocyte cell lines. The study highlighted the fact that the electrical properties of normal, pre-cancerous and cancerous oral keratinocytes are distinct.

During the same year, Henslee presented a study within which his group isolated target cell species from a heterogeneous sample of live cells with the help of contactless DEP. MCF10A, MCF7, and MDA-MB-231 human breast cells were employed to highlight different stages cancer; the study denoted that the trapping frequency was different from a type cell to another, while the voltage was swept from 20 to 30 V. MDA-MB-231 cells were successfully separated from population of MCF10A and MCF7 cells. One year later, Huang employed DEP to selectively concentrate cervical carcinoma cells (HeLa) from red blood cells. Cells positively charged were directed towards the centre of a DEP chamber as of high-electric-field. Moreover, it was demonstrated that the DEP force on HeLa cells was about seven times higher than the one acting on red blood cells. In about 2 minutes and 30 seconds, the cells of HeLa were concentrated with an electric field of 16V at 1MHz frequency.

Automatic microfluidic devices with improved capabilities provided by micropumps and microvalves facilitated the sorting performances of DEP system. DEP system providing a voltage of 15 Vpp at a frequency of 16 MHz was used to sort cells under a continuous flow.

Moreover, a 3D DEP chip device was developed with the aim of analysing the dielectric properties of osteosarcoma cells (MG-63 and SAOS-2) as well as an immune selected enriched skeletal stem cell fraction (STRO-1 positive cell). For all the lines under test, the dielectric properties demonstrated to be different, fact which convinced authors that sorting of specific cells from mixed populations can be further realized.

As a confirmation of DEP successful employment in living cells separation, sorting and analysis, a series of DEP devices were in the last years launched on the market. For example, Silicon Biosystem is commercializing an automated DEP system under the commercial name DEPArray which is allowing manipulation and isolation of single cell combined with imaging analysis. ApoCell is proposing ApoStream, a DEP device capable to identify circulating tumor cells (CTCs) and rare circulating cells based on dielectric behaviour and properties. The uptake of DEPIM (Dielectrophoretic Impedance Measurement) is demonstrated by the technology commercialized by Panasonic under the name Bacterial Counter which is allowing the rapid analysis of oral bacteria. Besides these, other DEP instruments may be mentioned, as 3DEP 3D Dielectrophoresis Cell Analysis System from Labtech, IG-1000 Plus Single Nano Particle Size Analyzer from Shimadzu or Ace Platform from Biological Dynamics.

IV.4 Hands-on design and fabrication of dielectrophoretic screen printed devices

The manufacturing of any dielectrophoretic printed devices is starting with the design and modelling. Within the first stage of manufacturing, the prototyping, the overall dimensions of the microelectrodes should be calculated and evaluated with the help of specialized CAD (Computer Aided Design) software. In the case of interdigitated microelectrodes (IDE), the number and the dimensions of the fingers as well as the distances (or gaps) between two fingers is fundamental in generation of an effective electric field gradient capable of trapping or sorting the target objects.

Overall, the efficiency of the devices based on interdigitated microelectrodes (IDE) is directly linked to the topology and dimension of the interdigitated microelectrode (IDE). The *efficiency, η,* is defined as the ratio between the sensing area and the overall area of the device:

$$\eta = \frac{sensing\ area\ S}{overall\ IDE\ area}$$

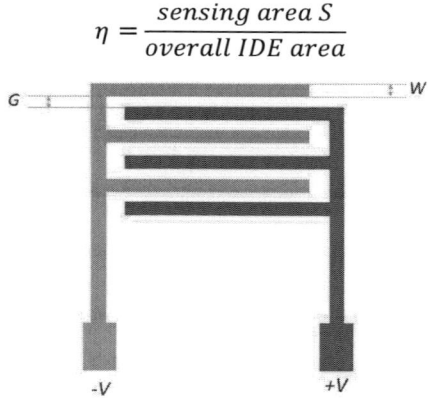

Figure 4.8 – Basic IDE structure

The efficiency is an adimensional unit and can be expressed much more rigorously in the case of an IDE as:

$$\eta = \frac{W}{W + G}$$

where *W* is the width of the finger, and *G* is the gap between two fingers.

The overall efficiency of the IDE when employed for dielectrophoretic trapping or sorting the field gradient is fundamental and can be enhanced through numerical simulations through finite element method. Moreover, the numerical simulations are allowing the modelling of substrate electrical behavior. Thus, the dielectric properties of the polymeric matrix can be analyzed on the basis of real permittivity (dielectric constant) and imaginary permittivity (dissipation or loss factor). In the case of our nanocomposite matrix the two dielectric parameters are revealing a behavior characteristic to strong polar materials. The dielectric constant is reaching a value of 3.51 at 10kHz and is decreasing slightly to 3.5 while increasing the frequency up to 125kHz and 3.494 while attaining 250kHz. Simultaneously, the loss factor is increasing from 0.002 at 10kHz up to 0.012 at 125kHz. Afterwards, the losses are decreasing down to 0.01 at 250kHz.

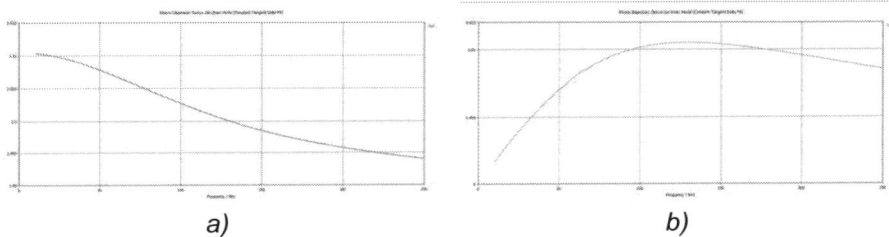

a)   b)

Figure 4.9 – Numerical simulation of (a) real permittivity and (b) loss factor evolution against the frequency

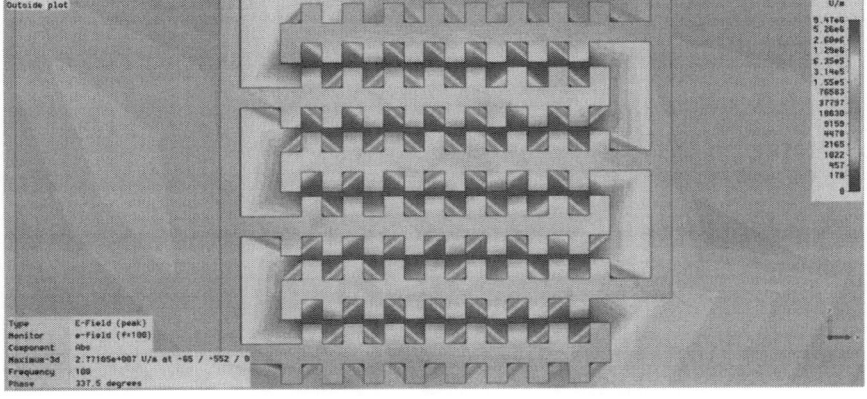

Figure 4.10 – Distribution of the electric field at the level of IDE with rectangular castellated microelectrodes

An exemplification of the electric field distribution at the level of interdigitated microelectrodes structure within a frequency domain from 100kHz is presented above.

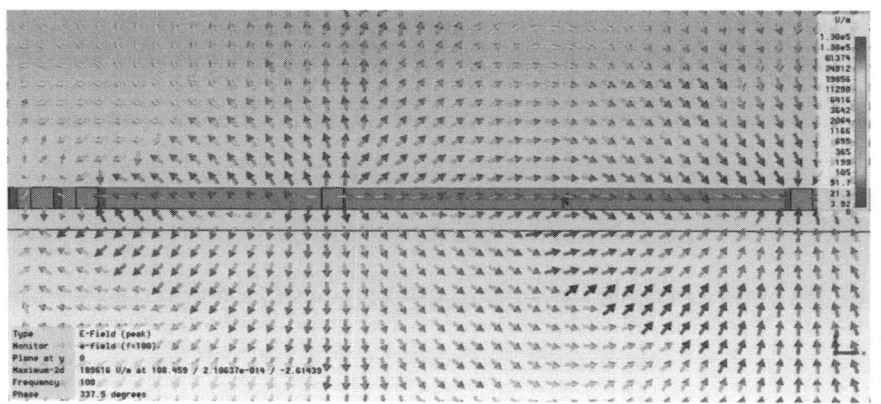

Figure 4.11 – Vector representation of the electric field at the level of IDE with rectangular castellated microelectrodes

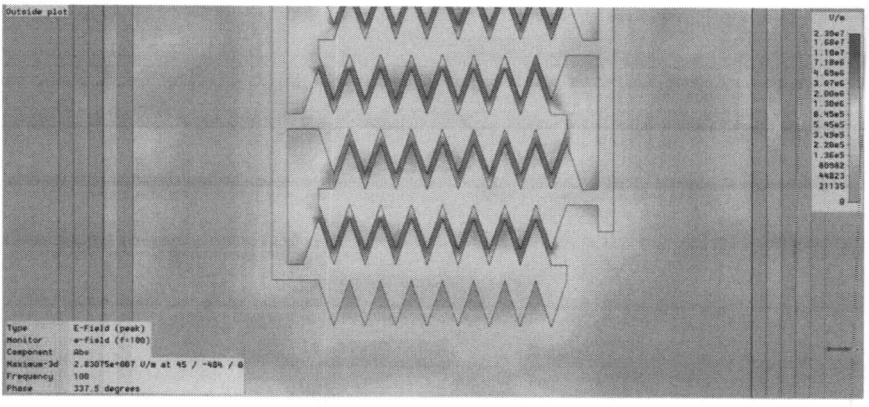

Figure 4.12 – Distribution of the electric field at the level of IDE with saw-shaped fingers

In the case of IDE structure with castellated fingers, the highest values of the electric field ($2.77*10^7$V/m) are reached at the level of two castellated structures, practically in the area where the distance between two fingers is the smallest.

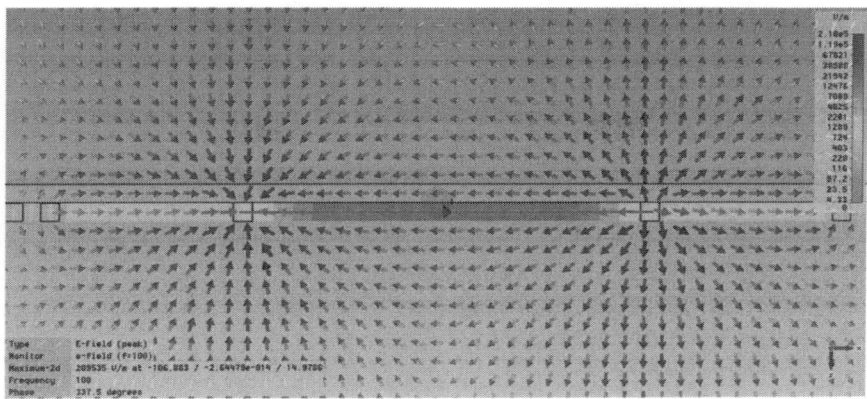

Figure 4.13 – Vector representation of the electric field at the level of IDE with saw-shaped microelectrodes

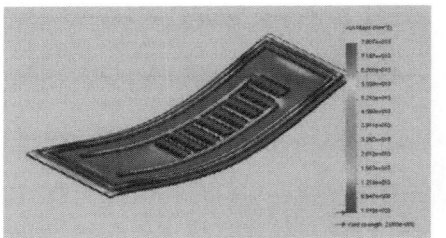
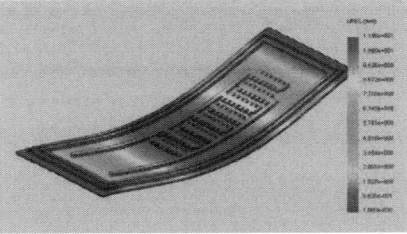

Figure 4.14 – Numerical evaluation of mechanical behaviour according to von Mises yield criterion – IDE with castellated fingers

Figure 4.15 – Numerical evaluation of mechanical behaviour according to von Mises yield criterion – classical IDE configuration

However, as reported by the literature, the electric field is reaching higher values at the level of fingers which are in the proximity of the terminals $(3.15 \times 10^5 V/m)$.

As a comparison, the distribution of the electric field at the level of interdigitated microelectrodes with saw-shaped fingers is provided as well. The electric field in this case is reaching values of the same order of magnitude as in the case of rectangular castellated IDE.

Besides the electric field distribution another essential issue of the printed device may be the mechanical problems which in the case of polymers are strong dependent on temperature. However, in spite of the fact that the mechanical failure of polymers is difficult to realize numerical modeling can be employed for evaluating equivalent mechanical resistance.

After assessing and establishing the final configuration of the printed patterns the following stage is the one of developing the screen mask containing the screen printing stencil.

The design of the stencil is developed once again in CAD applications; the dimensions of the design are traced at 1:1 scale with a line weight of 0.00mm. In the design below various designs of the IDE microelectrodes are presented, some of them being appropriate for sensing device development and dielectrophoretic manipulation of target particles, others are convenient for being employed in performing electrochemical impedance tests.

The following stage in screen printing is the stencil design development at the level of the screen mesh. Thus, the mesh is firstly cleaned with degreasing agent which is sprayed and eventual washed with the sponge. The degreasing is crucial before starting the loading of the screen with photo-emulsion. The degreasing solution should be prepared with a concentration of 1:25. (e.g. 250ml of degreaser in 1l of water). After spraying and washing the mesh should be cleaned entirely and dried completely in air or with a heater.

The coating of the mesh should be realized within a dark room under yellow light in order to avoid the exposure of the photo-emulsion at normal light.

The photo-emulsion coating process is to be done only with the sharp side of the coater with the mask frame in the vertical position in order to ensure uniform distribution of the photo-emulsion on the mesh. The coater should be moved from the down side of the mask frame to the up with a very smooth movement for ensuring constant thickness of the photo-emulsion on the mask. The process should be repeated twice on the side of the mesh which will be in contact with the printing substrate and once on the side of the squeegee. A heater may be used for ensuring a rapid drying of the photo-emulsion layer on the mesh. However, during photo-emulsion's drying process a temperature of 40ºC should not be exceeded. The photo-emulsion which remains unused on the coater may be recovered and restored. The humidity of the air in the room where the coating is done should be carefully controlled as the photo-emulsion is a highly hygroscopic material. Afterwards, the exposure foil printed with the layer design should be placed on the mask (the used of double sided adhesive tape might be useful). In case of a UV light source of 125W should be placed at a distanced sufficiently close to the mesh (about 50cm) and the overall exposure should be done for periods up to 45 min. While increasing the power of the UV light the exposure period should be decreased as well.

After removing the exposure foil the unexposed photo-emulsion will be removed from the mesh through a simple cleaning under water jet and thus on the mesh the wanted stencil is to be developed. The washing and drying of the photo-emulsion layer should be realized on both sides.

The screen printing mask can be realized as presented above but in order to ensure a better resolution of the stencil design it is preferably to subcontract its development to a specialized manufacturer. In case of a professional mask manufacturing process, the dimensions of the printing lines (openings within the photo-emulsions) may be decreased.

Besides, the photo-emulsion layer which is developed as of the fact that it will be realized with the help of (semi-) automatic coating equipment will have a uniform thickness which can be also rigorously controlled. Below, the mask developed for us by a specialized mask manufacturer is presented. The design of the patterns was developed in our laboratories but is in accordance with the designs employed by other IDE manufacturers but through deposition techniques and not through screen printing.

The screen mask presented below was developed and used for microelectrodes printing. It had the following characteristics:

Mesh type: 325MFT

String diameter: 13μm

Opening: 46μm

Thickness of gauze opening: 20μm

Space rate: 61% (((Width between two strings)^2/((String diameter + Width between two strings))^2)/100

Mask frame exterior dimension: 32*32cm

Mask frame exterior dimension: 28*28cm

Emulsion coating: 5μm

Number of microelectrodes: 50 per mask

a)   b)

Figure 4.16 – Image of the screen mask: a) the side of the mask in contact with the printing substrate, b) the side of the mask in contact with the squeegee

A fundamental element of the screen printing process is the formulation of a convenient ink or paste. The formulation, as presented within other chapters, is directly influenced by the morphology of the incorporated particles, surfactants utilization and particle density which is allowing ink's conductivity optimization.

Figure 4.17 – Manual screen printer

The final properties of the ink can be evaluated on the basis of the solid content, viscosity, particles' dimensions or Hegman gauge, resistivity after drying and drying time or drying process.

A fundamental element of the screen printing process is the formulation of a convenient ink or paste. The formulation, as presented within other chapters, is directly influenced by the morphology of the incorporated particles, surfactants utilization and particle density which is allowing ink's conductivity optimization.

The final properties of the ink can be evaluated on the basis of the solid content, viscosity, particles' dimensions or Hegman gauge, resistivity after drying and drying time or drying process.

In developing our IDEs we employed silver based ink provided by Applied Inks with the following technical specifications, according to producers technical sheet:

Shear Rate 20@25°C:	4300 – 5250 cps
Drying schedule:	90 seconds to 3 minutes @130°C
Total % NV Solids	71% ± 2%
Hegman Gauge	<25.0 µ
Surface Resistivity	< .025 Ω/square/mil

After printing and drying on polymeric substrate, the IDEs demonstrated very good adhesion to the substrate and very good electrical properties. However, as we forced the technical limits of the screen mask, the IDEs printed presented a series of defects which, as the number of fingers was sufficiently high, did not influenced the printed platform functionality.

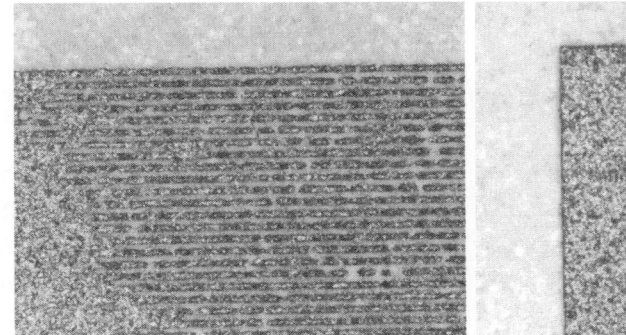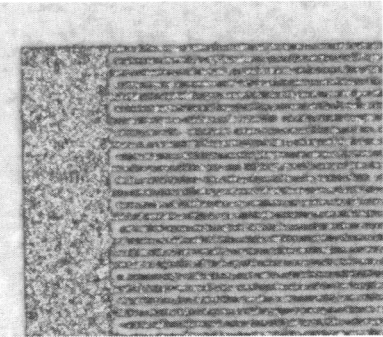

Figure 4.18 (a) – Silver based IDEs screen printed (By courtesy of Arcire Alexandru)

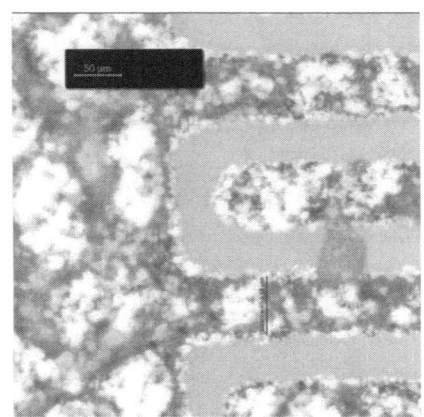

Figure 4.18 (b) – Detailed picture of silver based IDEs screen printed (By courtesy of Arcire Alexandru)

The interdigitated microelectrodes developed are critical for electromanipulation of nanoparticles or biological cells with the help of dielectrophoresis (DEP).

The dielectrophoresis bench employed is encompassing standard equipment for this type of laboratory set-up as function generator, peristaltic pump, dielectrophoretic micro-chamber, reservoir for the solution, microscope a d external light source.

The Perimax peristaltic pump employed was provided with 6 channels, adjustable flow control from 0.004ml/min up to 18ml/min and a speed range 1 – 33 rpm. The function generator with a frequency range from 50 mHz to 10 MHz, an output voltage: up to 10 V_{pp} (into 50 Ω), a distortion factor < 0.5 % up to 1 MHz and a rise and fall time: typ. 15 ns. The dielectrophoretic micro-chamber was realized of PDMS with a volume of 0.3ml and provided with an inlet and an outlet. The maximum dimension of the platforms with IDE was 9*25mm.

For imagistic monitoring of the dielectrophoretic process a Microscope Optika SZM-2 stereo was employed and provided with trinocular head, maximum magnification 45X, with zoom magnification achromatic objectives for bright field, suitable for life and materials science

For solution elaboration a magnetic mini-stirrer for mixing quantities up to 250 ml and a 500 - 1800 rpm speed range was employed.

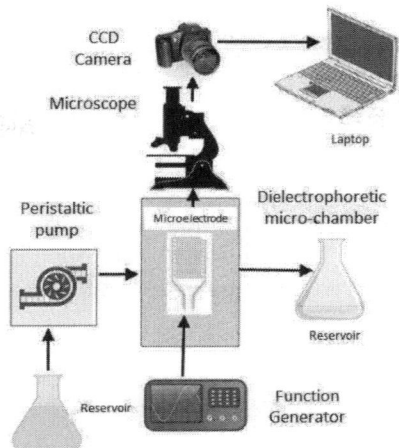

Figure 4.19 – Basic dielectrophoretic set-up (designed with the help of Arcire Alexandru)

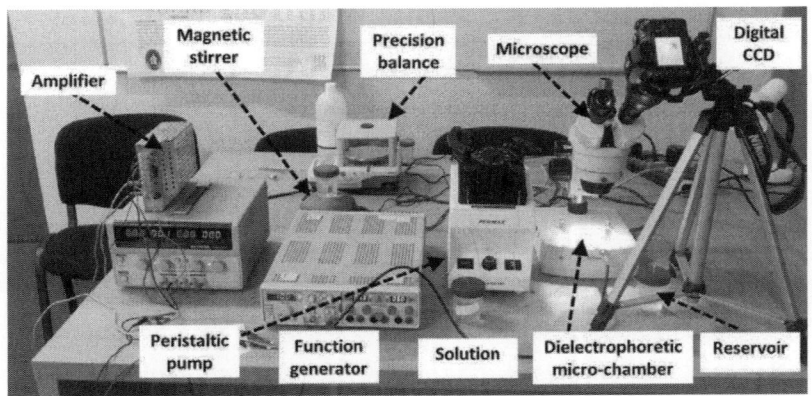

Figure 4.20 – DEP experimental laboratory bench (designed with the help of Arcire Alexndru)

In order to be electromanipulated with the help of dielectrophoresis, the fullerenes should be firstly dispersed within a fluid. Thus, 5 ethanol solutions containing various quantities of fullerenes as shown within the table below. The suspensions were realized at 1800rpm for a period of 60 minutes fact which allowed the fullerenes' unbundling. Afterwards, a quantity of 50ml of solution containing fullerenes was poured in a reservoir connected through a micro-pipeline to the inlet of the dielectrophoretic micro-chamber.

After introducing the platform with the IDE in the micro-chamber the peristaltic pump was switched and the solution started to flow. Simultaneously an AC field was switched on with the help of a function generator of $10V_{pp}$ and a frequency of 100kHz.

The trapping of the fullerenes started immediately after applying the dielectrophoretic force. During the process we also monitored the resistance of the electrodes as a confirmation of fullerenes bridges development. The resistance practically it is starting to decrease in the moment that the fullerenes bridges are formed. However, the amplitude of the AC field and the frequency can be varied in order to increase the quantity of fullerenes trapped.

Bibliography

[1] Pohl H.A, The Motion and Precipitation of Suspensoids in Divergent Electric Fields, J. Appl. Phys. 22, 869 (1951); doi: 10.1063/1.1700065

[2] Varshney M., Li Y., Interdigitated array microelectrodes based impedance biosensors for detection of bacterial cells, Biosensors and Bioelectronics 24 (2009) 2951–2960, 10.1016/j.bios.2008.10.001

[3] Wee W.H., Li Z., Hu J., Kadri N.A., Xu F., Li F. and Pingguan-Murphy B., Fabrication of dielectrophoretic microfluidic chips using a facile screen-printing technique for microparticle trapping, J. Micromech. Microeng. 25 (2015) 105015 (8pp), doi:10.1088/0960-1317/25/10/105015

[4] Gupta V. Jafferji I., Garza M. Melnikova V.O., Hasegawa D.K., Pethig R., Davis D.W., ApoStream™, a new dielectrophoretic device for antibody independent isolation and recovery of viable cancer cells from blood, Biomicrofluidics. 2012 Jun; 6(2): 024133, 10.1063/1.4731647

[5] S. Kinio and J. K. Mills, "Design of electrode topologies for dielectrophoresis through the use of genetic optimization with COMSOL Multiphysics," 2015 IEEE International Conference on Mechatronics and Automation (ICMA), Beijing, 2015, pp. 1019-1024. doi: 10.1109/ICMA.2015.7237625

[6] Proceedings Volume 6592, Bioengineered and Bioinspired Systems III; 65920O (2007); doi: 10.1117/12.724123

[7] Yafouz, B., Kadri, N. A., & Ibrahim, F. (2014). Dielectrophoretic Manipulation and Separation of Microparticles Using Microarray Dot Electrodes. Sensors (Basel, Switzerland), 14(4), 6356–6369.Sensors 2013, 13(7), 9029-9046; doi:10.3390/s130709029

[8] Suehiro, J., Ohtsubo, A., Hatano, T., & Hara, M. (2006). Selective detection of bacteria by a dielectrophoretic impedance measurement method using an antibody-immobilized electrode chip. Sensors and Actuators, B: Chemical, 119(1), 319-326. DOI: 10.1016/j.snb.2005.12.027.

[9] Hamada R., Suehiro J, Nakano M, Kikutani T, Konishi K., Development of rapid oral bacteria detection apparatus based on dielectrophoretic impedance measurement method, IET Nanobiotechnol. 2011 Jun; 5(2):25-31. doi: 10.1049/iet-nbt.2010.0011.

[10] Nakano M, Obara R. Ding Z.; Suehiro J., Detection of norovirus and rotavirus by dielectrophoretic impedance measurement, 2014, 10.1109/ICSensT.2013.6727678.

[11] Y. Huang, Biophys. J., 73 Aug. 1997, 1118-29

[12] Cemažar J, Kotnik T., Dielectrophoretic field-flow fractionation of electroporated cells, Electrophoresis. 2012 Sep;33(18):2867-74, doi: 10.1002/elps.201200265.

[13] Wang XB, Yang J, Huang Y, Vykoukal J, Becker FF, Gascoyne PR., Cell separation by dielectrophoretic field-flow-fractionation, Anal Chem. 2000 Feb 15;72(4):832-9.

[14] L. Wang (ed.), Advances in Transport Phenomena 2011, Advances in Transport Phenomena 3, DOI: 10.1007/978-3-319-01793-8_2, Springer 2014.

[15] Regtmeier J., Eichhorn R., Viefhues M., Bogunovic L., Anselmetti D., Review Electrodeless dielectrophoresis for bioanalysis: Theory, devices and applications, Electrophoresis 2011, 32, 2253–2273.

[16] Chiou C-H., Liang-Ju Chien L-J, Kuo J-K, Nanoconstriction-based electrodeless dielectrophoresis chip for nanoparticle and protein preconcentration, Appl. Phys. Express 8 085201, 2015.

[17] C.-F. Chou and F. Zenhausern, IEEE Eng. Med. Biol. Mag. 22, 62 (2003).

[18] N.Swami, C.F. Chou, V. Ramamurthy, V.Chaurey, Lab Chip 9, 3212 (2009).

[19] Chou C-F, Tegenfeldt J.O, Bakajin O., Chan S.S., Cox E.C., Darnton N., Duke T., Austin R.H, Electrodeless dielectrophoresis of single- and double-stranded DNA, Biophys J. 2002 Oct; 83(4): 2170–2179, doi: 10.1016/S0006-3495(02)73977-5.

[20] Qian C., Huang H., Chen L., Li X., Ge Z., Chen T., Yang Z. and Sun L., Dielectrophoresis for Bioparticle Manipulation, Int. J. Mol. Sci. 2014, 15, 18281-18309; doi: 10.3390/ijms151018281.

[21] Pethig R., Review—Where Is Dielectrophoresis (DEP) Going? ,Journal of The Electrochemical Society, 164 (5) B3049-B3055 (2017)

[22] Wang X -B, Huang Y., Becker F.F. and Gascoyne P. R. C., A unified theory of dielectrophoresis and travelling wave dielectrophoresis, J. Phys. D: Appl. Phys. 27 1571, 1994.

[23] Hagedorn R., Fuhr G., Müller T., Gimsa J., Traveling-wave dielectrophoresis of microparticles, Electrophoresis, Volume 13, Issue 1, Pages 49–54, 1992.

[24] van den Driesche S., Rao V., Puchberger-Enengl D., Witarski W.,. Vellekoop M.J., Continuous cell from cell separation by traveling wave Dielectrophoresis, Sensors and Actuators B: Chemical, Volume 170, 31 July 2012, Pages 207-214.

[25] Shafiee H., Caldwell, J.L., Sano M.B., Davalos R.V., Contactless dielectrophoresis: a new technique for cell manipulation, Biomed Microdevices, 2009 Oct;11(5):997-1006, DOI 10.1007/s10544-009-9317-5

[26] Cemazar J., Douglas T.A., Schmelz E.M., Rafael V. D., Enhanced contactless dielectrophoresis enrichment and isolation platform via cellscale microstructures, BiomMultilayer contactless dielectrophoresis: theoretical considerations, icrofluidics 10, 014109 (2016); doi: 10.1063/1.4939947.

[27] Sano M.B., Salmanzadeh A., Davalos R.V., Electrophoresis. 2012 Jul; 33(13):1938-46. doi: 10.1002/elps.201100677.

[28] Tzu-Keng Chiu, A-Ching Chao, Wen-Pin Chou, Chia-Jung Liao, Hung-Ming Wang, Jyun-Huan Chang, Ping-Hei Chen, Min-Hsien Wu, Optically-induced-dielectrophoresis (ODEP)-based cell manipulation in a microfluidic system for high-purity isolation of integral circulating tumor cell (CTC) clusters based on their size characteristics, Sensors and Actuators B: Chemical, Volume 258, 2018, Pages 1161-1173, ISSN 0925-4005, 10.1016/j.snb.2017.12.003.

[29] Khoshmanesh K, Nahavandi S, Baratchi S, Mitchell A, Kalantar-zadeh K. Dielectrophoretic platforms for bio-microfluidic systems. Biosensors and bioelectronics 2011;26:1800-14.

[30] Zhang C, Khoshmanesh K, Mitchell A, Kalantar-Zadeh K. Dielectrophoresis for manipulation of micro/nano particles in microfluidic systems. Analytical and bioanalytical chemistry 2010;396:401-20.

[31] Fatoyinbo HO, Hoettges KF, Hughes MP. Rapid-on-chip determination of dielectric properties of biological cells using imaging techniques in a dielectrophoresis dot microsystem. Electrophoresis 2008;29:3-10.

[32] Li H, Bashir R. Dielectrophoretic separation and manipulation of live and heat-treated cells of Listeria on microfabricated devices with interdigitated electrodes. Sensors and Actuators B: Chemical 2002;86:215-21.

[33] Bessette PH, Hu X, Soh HT, Daugherty PS. Microfluidic library screening for mapping antibody epitopes. Analytical chemistry 2007;79:2174-8.

[34] Pethig R. Review article—dielectrophoresis: status of the theory, technology, and applications. Biomicrofluidics 2010;4:022811.

[35] Pethig R, Menachery A, Pells S, De Sousa P. Dielectrophoresis: a review of applications for stem cell research. Journal of Biomedicine and Biotechnology 2010;2010.

[36] Zou H, Mellon S, Syms R, Tanner K. 2-dimensional MEMS dielectrophoresis device for osteoblast cell stimulation. Biomedical microdevices 2006;8:353-9.

[37] Hunt T, Westervelt R. Dielectrophoresis tweezers for single cell manipulation. Biomedical microdevices 2006;8:227-30.

[38] Cen EG, Dalton C, Li Y, Adamia S, Pilarski LM, Kaler KV. A combined dielectrophoresis, traveling wave dielectrophoresis and electrorotation microchip for the manipulation and characterization of human malignant cells. Journal of microbiological methods 2004;58:387-401.

[39] Lapizco-Encinas BH, Simmons BA, Cummings EB, Fintschenko Y. Insulator-based dielectrophoresis for the selective concentration and separation of live bacteria in water. Electrophoresis 2004;25:1695-704.

[40] Docoslis A, Tercero Espinoza LA, Zhang B, Cheng L-L, Israel BA, Alexandridis P, et al. Using nonuniform electric fields to accelerate the transport of viruses to surfaces from media of physiological ionic strength. Langmuir 2007;23:3840-8.

[41] Thomas RS, Morgan H, Green NG. Negative DEP traps for single cell immobilisation. Lab on a Chip 2009;9:1534-40.

[42] An J, Lee J, Lee SH, Park J, Kim B. Separation of malignant human breast cancer epithelial cells from healthy epithelial cells using an advanced dielectrophoresis-activated cell sorter (DACS). Analytical and bioanalytical chemistry 2009;394:801-9.

[43] Chen J, Li J, Sun Y. Microfluidic approaches for cancer cell detection, characterization, and separation. Lab on a Chip 2012;12:1753-67.

[44] Broche LM, Hoettges KF, Ogin SL, Kass GE, Hughes MP. Rapid, automated measurement of dielectrophoretic forces using DEP-activated microwells. Electrophoresis 2011;32:2393-9.

[45] Mulhall H, Labeed F, Kazmi B, Costea D, Hughes M, Lewis M. Cancer, pre-cancer and normal oral cells distinguished by dielectrophoresis. Analytical and bioanalytical chemistry 2011;401:2455-63.

[46] Yang L, Arias LR, Lane TS, Yancey MD, Mamouni J. Real-time electrical impedance-based measurement to distinguish oral cancer cells and non-cancer oral epithelial cells. Analytical and bioanalytical chemistry 2011;399:1823-33.

[47] Lu J, Huang C, Lull G, Jeon N, Monuki E, Flanagan L, et al. A NOVEL MICROFLUIDIC DEVICE COMBINING DIELECTROPHORESIS-BASED CELL PATTERNING AND 3D BIOMATERIALS.

[48] Wang X-B, Huang Y, Gascoyne PR, Becker FF, Hölzel R, Pethig R. Changes in Friend murine erythroleukaemia cell membranes during induced differentiation determined by electrorotation. Biochimica et Biophysica Acta (BBA)-Biomembranes 1994;1193:330-44.

[49] Henslee et al., Selective concentration of human cancer cells using contactless dielectrophoresis, DOI: 10.1002/elps.201100081

[50] Ching-Te Huang, Selectively Concentrating Cervical Carcinoma Cells from Red Blood Cells Utilizing Dielectrophoresis with Circular ITO Electrodes in Stepping Electric Fields, J. of Med. and Biological Engineering, 33(1): 51-58

[51] Tai et al., Automatic microfluidic platform for cell separation and nucleus collection, Biomed Microdevices, 2007 Aug;9(4):533-43.

[52] Ismail A et. al, Characterization of juman skeletal stem and bone cell populations using dielectrophoresis, J Tissue Eng Regen Med. 2012 Dec 6. doi: 10.1002/term.1629

[53] Meighan M et. al, Bioanalytical separations using electric field gradient techniques, Electrophoresis 2009, 30, 852–865

Druck:
Canon Deutschland Business Services GmbH
im Auftrag der KNV-Gruppe
Ferdinand-Jühlke-Str. 7
99095 Erfurt